Cellular Logic
Image Processing

CLIP was the WINNER
of the
BRITISH COMPUTER SOCIETY
TECHNICAL AWARD
for
1985

Cellular Logic
Image Processing

Edited by

M. J. B. DUFF

and

T. J. FOUNTAIN

*Department of Physics and Astronomy,
University College London*

1986

ACADEMIC PRESS

Harcourt Brace Jovanovich, Publishers

London Orlando San Diego New York
Austin Montreal Sydney Tokyo Toronto

ACADEMIC PRESS INC. (LONDON) LTD.
24/28 Oval Road
London NW1 7DX

United States Edition published by
ACADEMIC PRESS INC.
Orlando, Florida 32887

British Library Cataloguing in Publication Data

Cellular logic image processing.
 1. Image processing
 I. Duff, Michael J. B.   II. Fountain, T. J.
 006.4'2   TA1632

ISBN 0 12 223330 1

Printed in Great Britain

CONTRIBUTORS

K A CLARKE, IGE Medical Systems Ltd, Colney Street,
St Albans, Herts AL2 2ER.

M J B DUFF, Department of Physics and Astronomy, University
College London, Gower Street, London WC1E 6BT.

T J FOUNTAIN, Department of Physics and Astronomy,
University College London, Gower Street, London
WC1E 6BT.

H H-S IP, Imperial Cancer Research Fund Laboratories,
P O Box 123, Lincoln's Inn Fields, London WC2A 2PX.

G P OTTO, Department of Computer Science, University College
London, Gower Street, London WC1E 6BT.

S D PASS, Technology Group, Marconi Command and Control
Systems, Chobham Road, Frimley, Camberley, Surrey GU16
5PE.

D J POTTER, Department of Computer Science, Rensselaer
Polytechnic Institute, Troy, NY 12181, USA.

D E REYNOLDS, Cambridge Consultants Ltd, Science Park,
Milton Road, Cambridge CB4 4DW.

A M WOOD, Department of Physics and Astronomy, University
College London, Gower Street, London WC1E 6BT.

PREFACE

The Image Processing Group in the Department of Physics and Astronomy at University College London traces its origins to a project in 1958 which had as its objective the design and construction of automatic photoelectronic instruments for measurement of tracks in nuclear emulsions. In the following quarter of a century the Group widened its interest in image data to include all types of images and also carried out extensive studies of computer architectures designed specifically for the analysis of two-dimensional data arrays.

As in any other university research group, a large amount of the detailed work has been carried out by postgraduate students in research projects directed towards a doctorate, in this case under the supervision of one of the editors, Professor Michael Duff. However, construction of a full-scale computer involves a considerable engineering effort and the other editor of this volume, Dr Terry Fountain, was responsible for directing the construction programme for the image processing CLIP systems.

It seemed worthwhile to collect together a representative selection of extracts from the many PhD theses which have been written by members of the Group over the past years since these provide an insight into the philosophy which stimulated the research programme and led to the development of one of the world's fastest image analysis systems. Many others contributed to the research but space did not permit the inclusion of more material.

It is only right that our indebtedness to the late Professor Sir Harrie Massey, the previous Head of our Department, should be recorded here. Without his steadfast support the CLIP project would probably have not been able to survive, particularly during a period of very sparse funding. The continuing support from the present Head of

Department, Professor Franz Heymann, has ensured the project's continuation at a time when so many research programmes are facing extinction. Finally, the lasting confidence of the technical assessment committees advising the Science and Engineering Research Council, under whose sponsorship this project has been conducted, is gratefully appreciated.

No research work is carried out in a vacuum and an attempt has been made to acknowledge other research which has influenced the CLIP programme. In the area of processor array design, few projects have actually progressed to the stage of construction. The ICL Distributed Array Processor project was initiated in 1972 but construction of a pilot DAP did not start until 1975, two years after the commissioning of CLIP3. MPP, built for NASA by Goodyear Aerospace, was not completed until 1983. These and other competing systems now being designed have made their impact on the commercial world but only recently has a manufactured version of CLIP4 reached the market. The emphasis in the CLIP programme has been to build systems primarily in order to study the relationships between architectures and algorithms. Except for the ever-present restraints imposed by limited funding, engineering design for cost-effectiveness has not been of major concern; it is gratifying to see that colleagues in industry have been able to meet this challenge.

Our thanks are due to all those who have not only permitted us to use extracts from their theses but who have also not objected to the sometimes extensive editing which has been necessary in order to harmonise the extracts into what we hope is a coherent account. In this respect, the editors must accept responsibility for any errors and omissions which may have resulted as a consequence of their actions. We are particularly grateful to Miss Annette Harris for her careful and patient work in preparing the camera-ready material for this volume.

Michael J B Duff
Terry J Fountain

University College London
November 1985

CONTENTS

INTRODUCTION

It is, perhaps, one of the strange consequences of the manner in which television cameras are constructed that the traditional approach to image analysis has been not only to enter image data into a computer in raster-scan order, but also to process the data in a similar manner. The conventional computer program starts at the top left-hand pixel in the image and proceeds row by row through the image. More enlightened algorithms may divert from a completely regular path in order to follow around object contours but this can, and often does, lead to unwanted complexity.

In the late 1950s electronic computing was still in its infancy; university computers were few and far between and laboratory computers almost non-existent. In this discouraging environment, scientists were generating vast amounts of data so that the experimental bottleneck was often in the data reduction and data analysis phases of the project. The problem of how to cope with a 'data mountain' was probably most severely felt in High Energy Particle Physics where data appeared in the form of particle tracks in nuclear emulsions and cloud, bubble and spark chamber photographs. It was obvious that techniques would have to be developed that would speed up data processing by several orders of magnitude and various projects were initiated in association with particle physics laboratories in the United States, and in the United Kingdom and other European countries, especially in the CERN laboratories.

At University College London, in the (then) Department of Physics, a small research group was set up in collaboration with the already thriving Nuclear Emulsion Group. The new group, originally known as the Automatic Methods Group, was charged with the task of developing a range of devices which could be used to assist microscopists in their time-consuming work of finding and measuring the tracks of charged particles in their nuclear emulsions. The systems produced at that time usually comprised a microscope with one or more motorised movements (on the X-Y stage, on the objective or in the graticule eyepiece), together with some form of digital position encoder on each movement. Most devices were semi-automatic in that an observer was required to set the microscope measuring graticule on the desired part of the particle track. Calculations on the digitized data were performed either by solenoid-controlled mechanical adding machines or else by arithmetic-logic circuits constructed from solenoid-operated relays. As time progressed into the middle 1960s, the cumbersome and not always reliable relays were replaced by diodes and eventually by transistors and, in more advanced systems, the human microscopists could themselves be replaced, for a very limited set of tasks, by television cameras or scanned optical systems and photomultipliers.

Despite a reasonable degree of success in projects such as these [e.g. 1], it soon became apparent that the systems being produced were strictly limited in their capabilities. An inherent weakness ran through them all: although they were well able to remove much of the drudgery from straightforward measuring tasks, particularly on virtually straight tracks, complex images 'confused' them. Also, the systems could not of themselves be of help when it came to finding and recognising tracks in particular configurations or patterns. Systems designed to extract data from track chamber photographs were similarly restricted in their effectiveness, even though by this time large mainframe computers were appearing on the scene and bubble chamber data computations were being performed off-line on data collected from the semi-automatic measuring machines.

The particular difficulty affecting this method of working concerned the enormous proportion of irrelevant data in the photographic images. It was not practicable to record all the data in digital form, using subsequent computation to reject the unwanted portion. Major rejection before recording was essential and had to be the responsibility of human operators. What was needed was some form of built-in pattern recognition capability in the recording or scanning system itself.

At the same time as physicists were trying to find ways of analysing vast amounts of data, biologists and their engineering colleagues were attempting to understand the structure and behaviour of the nervous system by designing neural analogues. Neurons were modelled and assembled into networks (neural nets) which exhibited some of the observable characteristics of their physiological counterparts. Most of the model design was, of necessity, inspired guesswork since detailed anatomical and physiological information was not (and still is not) available. Nevertheless, the models helped to provide insight to guide further investigations of neural behaviour and, significantly from the point of view of computer science, served to inspire a new approach to data analysis.

Four influential papers were published in the late 1950s and early 1960s: two by Unger [2,3] describing a mesh of simple processors for image data analysis, one by Lettvin et al. [4] proposing a model for the visual system of the frog, and another by Hubel and Wiesel [5] presenting a now classic description of the structure and function of the cat's visual cortex. Herscher and Kelley [6] modelled the frog's visual system and Unger's ideas were embodied into designs for image analysis computers proposed by Slotnick et al. [7] and, later, McCormick [8]. The work by Hubel and Wiesel continues to direct attention to the use of clusters of detectors and processors as a means of implementing local neighbourhood operators for line segment detection and for spatial filtering.

At University College London, attention focussed on the specific problem of automatically locating areas of interest in charged particle tracks. Since a straight track can be assumed to be not interacting, it follows that interaction points are also points at which the track deviates so the problem reduces to one of finding sharp changes of direction in the track, positions which are usually referred to as vertex points. Once such positions could be found, measuring tracks radiating from each point was a process lending itself readily to automation. It was decided to build a parallel processing system, loosely modelled on the retina, to find vertex points automatically and rapidly, as a preprocessing stage prior to measurement.

UCPR1

In 1967 Duff et al. [9] demonstrated a fixed logic, 400-processor system arranged as a 20 x 20 square array. Charged particle track images were projected on to an input array of photodiodes. The outputs from the photodiodes, suitably thresholded, were then summed over three by three windows. A further layer formed sums over a five by five ring of outputs from the first summation.

Finally, a global threshold was applied, reducing from a maximum until at least one processor output exceeded the current threshold level. At this point, one or more miniature indicator lamps lit up in the corresponding positions

in the output array. These lamps indicated the approximate
position of a vertex in the input image. In a subsequent
development, the same system, operating with an increasing
rather than a decreasing threshold (and a small circuit
modification to ensure detected points were elements of the
input particle track image), was able to locate line ends.
By combining the two functions and by partitioning the output into regions, a crude line feature analyser was constructed enabling simple recognition of line figures (such
as a subset of the alphanumerics).

The capabilities of UCPR1 were extremely limited
although it did serve to illustrate the potential of parallel processing arrays. The similarity to the retina was
superficial so that the system could certainly not be
described as a retina model. Furthermore, as a device for
aiding the analysis of track chamber photographs, it was of
no direct value. What, therefore, was learned from this
project?

In the first place, the processing speed was impressive
by the standards of the day, vertices being found in about
20 ms. Secondly, it was soon clear that a simple four-
layer processing array (comprising input, two summations and
output) could hardly be applied to more than a few elementary tasks. Thirdly, the technology employed in the array
construction, which incorporated hand-assembled multilayer
printed circuit boards, was clumsy and inappropriate. The
possibility of constructing an array with, say, ten or more
layers of logic, each performing a fixed function but allowing data to be moved in either direction between layers, was
given serious consideration. However, the technical difficulty of making the necessary connections between the layers
was daunting and other lines of thought were pursued in
order to avoid this problem.

The Diode Array

In the Laboratorio di Cibernetica, near Naples, Italy,
Levialdi [10] had been investigating the processing capability of a simple array of switches, connected so as to detect
closed loops in binary figures. This array demonstrated
the power of propagation as a means of passing information
between processors over long distances in the array. Dis-
cussions between one of us (MJBD) and Professor Levialdi,
marking the beginning of what was to become a long and
fruitful collaborative association, resulted in a study
being carried out to determine the intrinsic processing

capability of an array of 'minimal' processing elements. These elements were to contain the smallest set of components that could be thought to have the required capability: an input and an output for image data, paths between adjacent processors, and a means for inverting or switching on or off the outputs from the processors, both the data outputs and the outputs to the neighbouring processors. A processing 'cell' was defined, incorporating a two-pole, double-throw toggle switch, a neon indicator (with a series resistor) and diodes linked to input and output buses surrounding each cell. The series resistor permitted the neon indicator to be illuminated or extinguished by applying a positive voltage to one or other side of the indicator when the resistor itself was earthed. The array was 'programmed' by choosing an appropriate set of connections within each cell, the connection pattern being identical in every cell.

A trial 5 x 5 cell array was constructed, using electromagnetic relays to select a variety of connection patterns, primarily to demonstrate visibly the ideas forming the basis of the project. At the same time, a Fortran simulation was written and linked to a Monte Carlo program which explored arbitrary connection schemes by choosing them at random and applying the resulting array to a standard test image. Schemes which generated new functions, as shown by the appearance of a previously unseen output image, were stored for future reference. The random search was moderated by a set of intuitively obvious constraints, such as rejection of schemes which would short out the power supply or would only connect to one side of the toggle switch. A typical result occurred when 3906 trials generated 73 schemes, each with different processing properties, the program taking about 30 minutes of execution time in an IBM 360 computer. The discovery curve was clearly flattening by the end of the run and it was interesting to note that all the ten schemes which had already been devised and wired into the demonstration array had emerged from the Monte Carlo search after 18 minutes of processing time.

The Diode Array served as a useful testbed for developing ideas about arrays of mesh-connected processors and pointed strongly to the potential value of propagation. Attempts to describe propagation algebraically met with the difficulty that the diode paths were bidirectional for signals of opposite polarity [11]. By specifying a processing element in terms of pure logic functions (rather than real circuit elements), this problem could be

circumvented.   It then only remained to translate the logic
specification into the then newly available small scale
integrated circuit components.

CLIP1

Despite encouraging results from simulations, little had
been done to investigate the practical problems that might
arise when mesh-connected processor structures were actually
constructed and operated.   A haunting fear for those
involved in the 400-processor UCPR1 had been that a short-
circuit in one element might propagate throughout the array,
leaving a trail of burnt-out components in its path.   With
the integrated circuit network about to be built, dubbed
CLIP (a Cellular Logic Image Processor), it was thought that
an added hazard might be that the array would go into oscil-
lation due to the presence of signal feedback paths through
chains of neighbouring elements.

The processing element circuit was extremely simple and
comprised eight two-input and two four-input NAND gates in
three small scale integrated circuit packages.   One hundred
of these elements were mounted on one large printed circuit
board to form a 10 x 10, 4-connected array (i.e. each element connected to its N, E, S and W neighbours).   A
flying-spot scanner generated a binary image which was then
stored in the input memory (a shift register).   One of
three possible functions was selected by appropriate control
lines and the contents of the shift register connected to
the array element inputs through 100 parallel output lines.
The output data was buffered before re-entry into the shift
register for display.   The three functions chosen for this
trial array were extraction of the contents of closed loops
(of 1-elements), extraction of sets of 1-elements connected
to the array border, and extraction of the outer edges of
objects composed of 1-elements.   Images were displayed by
modulating the brightness of a 10-line raster on a display
oscilloscope.   The input image could be altered pixel-by-
pixel using a light pen to interact with the display of the
input image stored in the shift register.

The CLIP1 project and the subsequent development of
CLIP2 and CLIP3 was the doctoral research programme of D M
Watson [12], working under the supervision of one of us
(MJBD).   The fears expressed above were largely allayed by
CLIP1, instilling confidence to proceed to a more general
processor.

CLIP2

There are only 16 Boolean functions of two Boolean variables and they can be generated by conditionally summing the four minterms. In CLIP2, a 16 x 12 hexagonally-connected array of processing elements, every element comprised two programmable Boolean processors each implementing independently the full set of functions of two inputs. One Boolean processor generated an output which it then transmitted to its six neighbours, whilst the other processor output into an image memory for subsequent further processing or display. The two inputs were either both binary images or else a binary image at one input and the ORed interconnection signals at the other. Additional circuits allowed a second image to be ORed in with the interconnection signals at the second input. Once again, the input image was generated by means of a flying-spot scanner in conjunction with a light pen. Instructions were entered manually (as 12-bit words) and stored in a 32-word memory. The system was completed in 1972.

Apart from the obviously small image size, CLIP2 suffered from one major shortcoming: its operations were completely non-directional. When information was sent from processor to processor, all neighbouring processors were addressed simultaneously so that directional information was not retained or transmitted. Possible functions were therefore limited to those in which direction was not implied, involving properties such as enclosure, expansion and connection. This extremely severe limitation was anticipated from the start; CLIP2 was still seen as a testbed for studying array control strategies and for exploring the propagation of signals through arrays. It is also a fact that, in the early days of the CLIP programme, these arrays were regarded as special-purpose image processors, not as general-purpose computers. It was, perhaps, the study of CLIP2 which suggested that a small amount of sophistication of the processing elements would generalise the array so that it would be inherently capable of performing any image operation.

CLIP3

The required processor enhancement appeared in CLIP3 in which the interconnection structure between processors was improved by individually gating each direction (now 6 or 8, providing hexagonal or square connectivity) and by replacing

the OR gate by a threshold gate.   Strictly speaking, the
threshold gate was not a necessity although it reduced substantially the number of instructions needed to implement
certain valuable local neighbourhood operations.   The
instruction word was increased to 24 bits and the instruction memory to 256 words.   Sixteen bits of local image
memory were provided at every processor and various modes
for loading data into this memory were allowed.

Many programs were written for CLIP3, ranging from conventional image processing to non-image operations such as
maze solving, electrostatic field calculation and lay plan-
ning (fitting together garment pattern parts so as to use
cloth economically).   The small image area limitation was
partially overcome by constructing a scanning system which
covered a 96 x 96 pixel image with the 16 x 12 processor
array.   This development made it feasible to write programs
to process grey-level images, using bit-serial algorithms;
the smaller image size had been too small to allow grey-
level images to be represented meaningfully.

The next phase in the CLIP programme was devoted to
refining the processor specification, including only those
parts which seemed essential but adding a few features which
were judged likely to produce a significant improvement in
performance.   The motivation was to reach a design which
could be fabricated as a large scale integrated circuit,
principally so that larger arrays could be built at a reasonable cost.   The vastly slower performance of the scanned
version of CLIP3 (about 3000 times slower than CLIP3 itself)
stimulated the desire to build a 96 x 96 processor array
which would recover most of the speed lost in the scanning
operation.

CLIP4

Despite the usefulness of the CLIP3 threshold gate, it
was an expensive luxury in terms of the number of transistors needed to implement it.   On the other hand, the need to
double the local memory was apparent and a useful increase
in processing power could be achieved at little cost by
incorporating the few gates needed to convert the processor
optionally into a full binary adder.   The only readily
available integrated circuit technology (available to the UK
university community, that is) was metal gate NMOS.   Preliminary calculations indicated that it should be possible
to design a circuit comprising eight processors, each with
32 bits of local memory, the resulting chip being of a size

which might be expected to give a satisfactory yield.   The
target price for the eight-processor packaged device was £6
(1974 prices).

Sadly, the initial optimism in this project was soon to
be dampened.   A full array of 96 x 96 CLIP4 circuits was
not forthcoming until early in 1980, some six and a half
years after the design contract was placed.   In the inter-
vening period, three design houses had applied themselves to
the project and, at the end, the circuits produced (usually
referred to as CLIP4A) could only be used at 40% of the
design clock frequency, 1 MHz rather than 2.5 MHz.
Undoubtedly, this shortcoming was the result of an attempt
to salvage the aborted earlier work when the last company
involved took the project to conclusion.   Subsequently, the
same organisation, Swindon Silicon Systems Ltd, carried out
a complete redesign from first principles and eventually
produced a version (CLIP4D) which fully met the original
specification.   In the Chapters which follow, the work
described relates to the CLIP4A circuit but applies equally
to CLIP4D in all respects other than performance times.

Details of the operation of the CLIP4 system appear in
subsequent Chapters.   The system comprises an array of 9216
bit-serial processors, configurable as either a square or
hexagonal 96 x 96 array, interfaced to two framestores holding 6-bit images.   The input store can be loaded from a
variety of television camera optical workstations via an A/D
converter which samples the central part of a non-interlaced
standard television frame, producing the required digital
image.   The output memory offloads through a D/A converter
for display on a standard television monitor, and facilities
are provided for mixing, at variable intensities, the original analogue image and one or other of the input or output
digital images.   The user communicates with CLIP4 via a DEC
PDP-11 computer operating under UNIX.   The serial host is
time-shared but CLIP4 itself can only accommodate one user
at a time.   However, control can be switched to another user
in a matter of seconds and the comprehensive file system
provided by UNIX enables users to keep track of the
whereabouts of both their programs and their image data.

In the following Chapters various aspects of the system
will be described by some of the postgraduate students who
have used it in the five years since it was commissioned.
Many other users have worked with the system and CLIP4,
although a prototype machine, has given excellent service.
Two similar machines have been constructed by the Science
and Engineering Research Council (Rutherford Appleton
Laboratory) so that others could gain experience of CLIP

processing.   Prior to this, some laboratories had written
CLIP4 emulators.   Strange to say, one of these emulators
was found to give substantial processing speed gains over
the serial machine on which it was running, indicating that
algorithm design can sometimes be improved even by thinking
about computer architecture!

More recently, a commercial organisation (Stonefield of
Horsham, England) has started marketing their own version of
CLIP4 so the community of CLIP users is now sharply increasing.   It is interesting and pleasing to see how ideas which
were once regarded as strange if not bizarre are now widely
accepted and being put to good practical use.

CHAPTER ONE

BASIC CLIP PROCESSING

Stephen Pass was the first postgraduate in the Image Processing Group to have the opportunity of implementing parallel processing algorithms on a working CLIP4 system. The title of his thesis 'Parallel Techniques for High Level Image Segmentation Using the CLIP4 Computer' reflects the importance attached to a thorough understanding of operations on the then new system. The section of thesis reproduced here serves as an overall introduction to the system hardware and software as well as to the basic operations which may be implemented thereon. The description is couched in terms of the CAP4 assembly language (see Chapter 3), the use of which emphasises the fundamental operations employed.

CELLULAR LOGIC
IMAGE PROCESSING
ISBN 0 12 223330 1

1.   CLIP4   HARDWARE

The heart of CLIP4 is the processing element (PE) shown in Figure 1.1.   An array of 96 x 96 such elements is connected as shown in Figure 1.2 and all processors in the array execute the same instruction simultaneously.   Each element contains a Boolean processor, one of whose inputs is the local image value in the A-register.   The other is derived from a combination of the neighbouring processor outputs and the value of the B-register.   For single image operations only the A-register is used, the B-register being left disabled, but for double image processes the B-register is loaded with the desired image and the enable-B line is set.   Double image processes fall into two distinct categories:

1. Non-propagating – all neighbourhood inputs are disabled and the output pattern is the point-wise Boolean combination of the two inputs.

2. Propagating – here data is passed between PEs. Either local or global propagating functions can be defined but the effects on the input image in the A-register are conditioned by the contents of the B-register.   Perhaps the most obvious example of such a process is labelling where a global propagating function allows the B-register to pick out (or label) certain objects in the A field.

The Boolean processor has two independent outputs, the new image value (D) and the neighbour output (N), which are each defined by four control lines giving a total of 256 possible transform functions.   Not all functions produce sensible image transforms, particularly those involving inverting global propagation which sometimes give rise to unstable output patterns.   By suitable choice of enabled neighbourhood directions inverting propagation can be used to generate regular dotted or striped patterns which have been found to be of limited use in forming test images.   Either square or hexagonal connectivity, with each neighbour input individually gated, can be specified under program control.

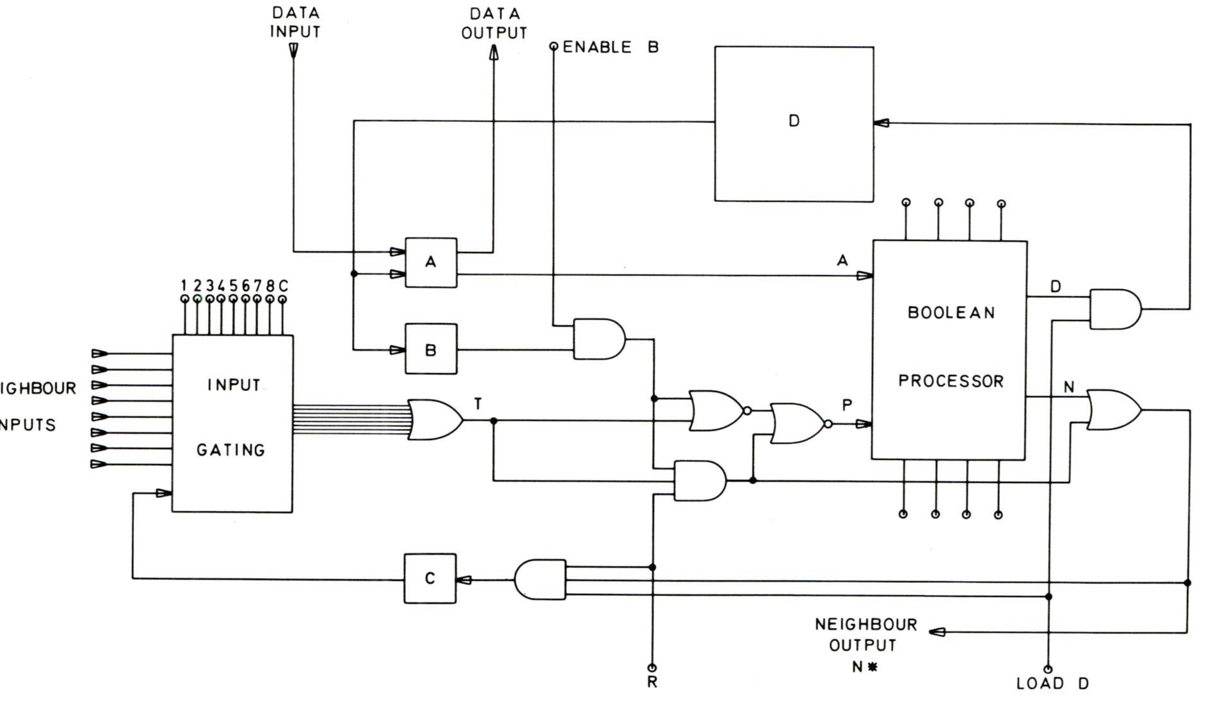

Figure 1.1   CLIP4 processing element.

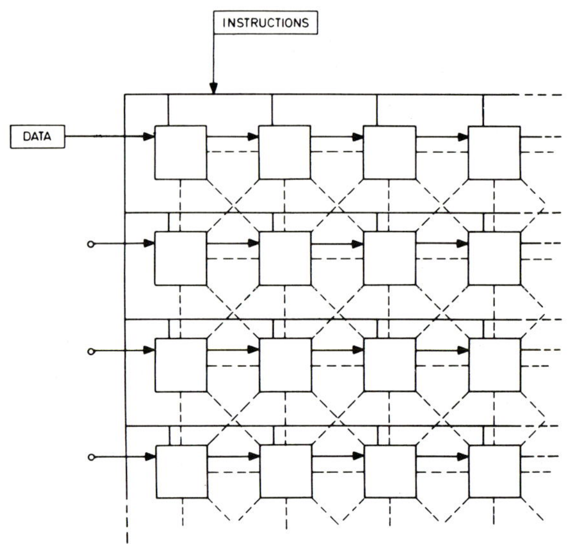

Figure 1.2   Connections in an array of processing elements.

To improve the efficiency of arithmetic processing a full-adder was incorporated within the PE.   When the R line is disabled the P input is just the Boolean OR of the B and T values and the PE performs normal pattern transformations, but when R is enabled the two NOR gates and the AND gate providing the P input to the Boolean processor form a half-adder.   The Boolean processor is set as another half-adder, giving a full-adder operation for the PE.   The result of the addition is formed at the D output and the new carry value comes from the N output.

The T value comes from one of two sources, depending on the mode of arithmetic processing.   For bit-column arithmetic, where the number is stored along a column of PEs, T comes from an adjacent PE (i.e. one neighbour input is enabled), but for bit-plane arithmetic, where the number is stored in one PE, T comes from the C-register.   This holds the carry resulting from the previous addition of corresponding bits of the operands, ready for combination with the next bits of the operands.

The local memory for each PE is only 32 bits, which is adequate for most binary image processing but is a severe restriction when processing grey-valued images. Low precision integer or fixed-point bit-plane arithmetic is practicable but multiplication and division quickly exhaust the available memory. Floating-point bit-plane arithmetic is, however, completely unfeasible. Bit-column arithmetic can be performed with much greater precision since each memory plane holds ninety-six 96-bit numbers. Fixed-point multiplication, division and square-rooting have been implemented in bit-column arithmetic on CLIP4 (addition and subtraction are each only one instruction).

A custom integrated circuit incorporating eight PEs in a 4 x 2 sub-matrix was commissioned for CLIP4. Local memory for the PEs is provided on-chip, making 256 bits of RAM for each integrated circuit. The circuit was originally expected to work with a basic clock rate of 2.5 MHz but problems with the design necessitated a downgrading to 1 MHz, a 60% reduction in processing speed. Four-phase NMOS technology was used for the chip design which yielded a fairly large circuit of 0.168 x 0.177 in of silicon, encapsulated in a 40-pin DIL package. 1152 such devices were required for the 96 x 96 array of CLIP4.

A schematic diagram of the overall CLIP4 system is shown in Figure 1.3. The analogue television signal is digitised into six bits with a parallel A/D converter which uses 62 comparator circuits. Each plane of the image is stored in a shift register 9216 bits long. These shift registers have proved moderately sensitive to temperature variations and it is proposed to replace them with static RAM circuits in future CLIP4s. From the input memory binary patterns of four types can be loaded into the processor array via the buffer memory. These patterns are:

1. Points brighter than a specified value (threshold above).

2. Points darker than a specified value (threshold below).

3. Points at the specified level.

4. Points from one plane of the image.

Thus, to transfer a full grey-value image into the array, six input operations have to be performed.

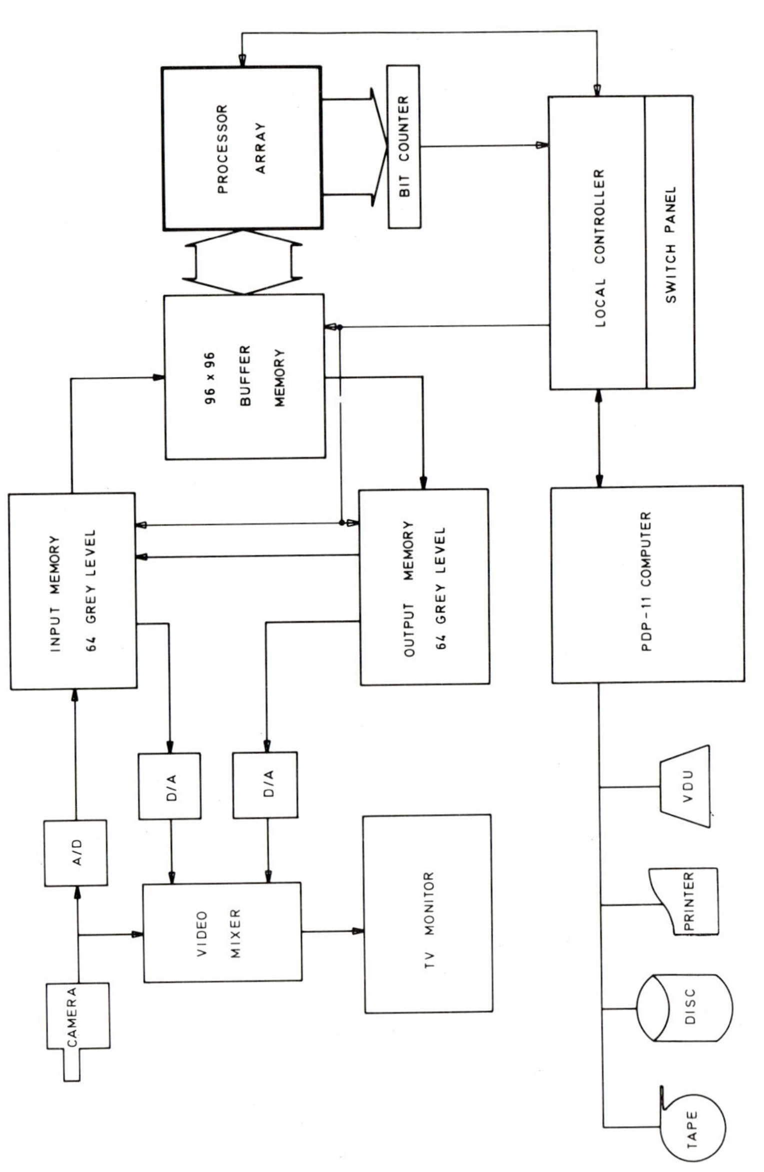

Figure 1.3   The CLIP4 system.

Only the central ninth (1/3 x 1/3) of the full
television picture is normally digitised and available to
the processor array, but a facility exists for quadrupling
the digitised area by sampling alternate lines of the raster
scanned image.   This can be likened to a coarse zoom facility.   For display, the processed picture is loaded into the
output memory, which is physically similar to the input
memory.   On the television monitor the contents of either
the input or the output memory can be displayed, together
with the surrounding analogue television picture.   It is
also possible to overlay the digitised image on the analogue
picture which has proved useful for relating a binary result
image to a grey input picture.

CLIP4 is closely interfaced to a DEC PDP-11 minicom-
puter.   The program memory of CLIP4 amounts to 32 Kbytes
and is within the address space of the PDP-11.   Programs
for execution by CLIP4 are assembled by the PDP-11 and
loaded into the appropriate place in the shared memory.
The PDP-11 sets a flag in a control register to start CLIP4
running and then waits for an interrupt which indicates that
CLIP4 has halted.   Concurrent processing by CLIP4 and the
PDP-11 is not possible since the shared memory would lead to
addressing conflicts between the two machines.

2. THE CLIP4 ASSEMBLY LANGUAGE

The assembly language of CLIP4 is called CAP4 (CLIP4
Assembly Program) and is fully described in Chapter 3.   A
CAP4 statement typically consists of a combination of four
possible fields:

LABEL: OPERATOR   OPERAND   ;COMMENT

The LABEL consists of up to six alphanumeric characters and
is only required by those statements to which a branch will
occur.   The OPERATOR can take the form of either an
instruction mnemonic or an assembler directive.   An
instruction mnemonic corresponds to executable machine code,
while a directive tells the assembler how to deal with a
section of program or with a set of characters.   The
OPERAND represents the data or address on which the OPERATOR
will act and, as such, its form is dependent on the OPERA-
TOR.   Not all OPERATORs have an associated OPERAND, its
presence being implied by the nature of the OPERATOR.   The
COMMENT is optional and allows annotation of programs.

Instruction mnemonics fall roughly into five categories:

1. Array
2. Register
3. Branches
4. Input/output
5. Miscellaneous

Of these mnemonic types, only the array instructions will be described at any length since these are fundamental to understanding how CLIP4 is programmed. The register instructions are essentially limited to addition and subtraction and are mainly used in calculating the addresses of operands for array instructions and in counting for program loops. The branch category covers unconditional, conditional and subroutine jumps, all of which are used in controlling the order of execution of a program. There are conditional branches for the state of sense-switches mounted on the control panel, the result of a register operation or the result of an image processing operation. This latter feature enables decisions to be made on the basis of the effect of an image transform. The input/output instructions, as their name implies, control the I/O system and fulfil the operations briefly described in the previous section on hardware. Perhaps the most important instruction in the miscellaneous category is COUNT which finds the number of 1 bits in a binary image. The number is transferred to a register and gives a numerical value to the result of an image processing operation.

Returning to the array instructions, there are four types. These are LDA, LDB, SET and PST. Referring to Figure 1.1, LDA simply loads the A-register of a PE with a single bit of data from a location in the D memory specified in the operand field, e.g. LDA 10 copies the contents of memory location (D-level) 10 into the A-register. Every PE performs this instruction at the same time. Similarly, LDB loads the B-register from a specified memory location.

The SET instruction is the basis of array processing on CLIP4, since it specifies the manner in which the array will operate during a processing cycle. It consists of three subfields which are, in order, the output definition, the propagation definition and default-options.

The output definition consists of a Boolean function of the A (local image value) and P (propagation from adjacent cells) inputs which are combined to produce the D (transformed image) output. The propagation definition

specifies, in a direction field, from which adjacent cells
an input will be received, and also defines a Boolean function of the A and P inputs which are combined to form the N
(neighbour) output.   Included within the direction field
are the B and C options which respectively enable the B- and
C-registers to contribute to the propagation input.

The default-options allow the array to be used with hexagonal instead of square tessellation, the outer edges of
the array to provide a propagation value of 1 rather than 0,
and the R line to be enabled for arithmetic processing.   A
typical instruction would be:

$$
\begin{array}{llcll}
 & & \text{direction field} & & \\
 & & \text{specifying} & & \\
 & & \text{accepted} & & \text{default-} \\
 & & \text{directions} & & \text{option} \\
\text{SET} & \text{A+-P} & [2] & \text{-A.P,} & \text{E} \\
 & \text{D output} & & \text{N output} & \\
 & \text{definition} & & \text{definition} & \\
\end{array}
$$

where the logical operations are:

    + = OR
    . = AND
    - = NOT
    @ = EXCLUSIVE OR

The numerical directions surrounding a PE are shown in Figure 1.4, and the E default-option sets the edges of the
array to logical 1.

The PST instruction is a mnemonic for Process and STore.
When this is executed the array goes through a processing
cycle.   The data loaded in the A and B-registers is processed according to the functions defined in the previous
SET instruction and the result is stored in the memory location specified in the operand field of the PST, e.g. PST 5
would store the result in D-level 5.   A typical sequence of
CAP4 statements would be:

```
LDA 2
SET P,[8]A
PST 2
```

The result would be a shift to the right by one pixel of a
binary image held in D-level 2.   If the image is considered
to consist of white objects (sets of 1s) on a black

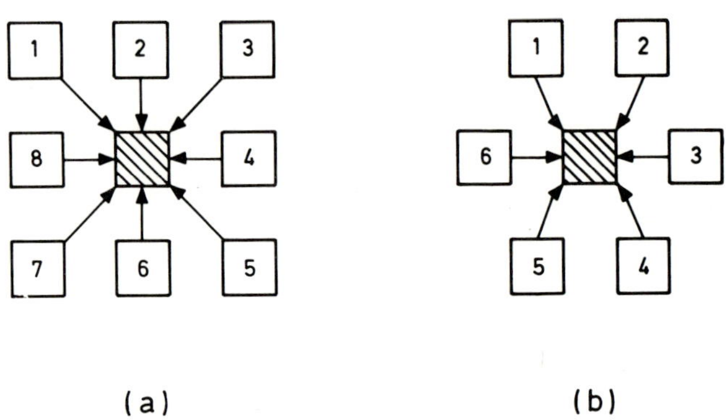

(a)   (b)

Figure 1.4   (a) Square and (b) hexagonal surround directions.

background (0s) then the propagation definition specifies that the value passed to adjacent PEs is the local image value. However, since only direction 8 is enabled then a PE only receives a propagation signal from the PE on its left. The output definition is P and so the D output takes on the value of the propagation input. Thus, each PE passes its own value to the right and takes on a new value from the left, the overall effect being a right-shift of the image.

3. BASIC OPERATIONS ON BINARY IMAGES

The close relationship between the architecture of CLIP4 and the structure of an image allows several important image transforms to be performed in one machine operation. These transforms are dependent on the interchange of information between pixels and this is easily achieved with the cellular structure and local connectivity of the processor array. The basic image transforms can be broadly divided into two categories. The first depends on each pixel's relationship with its immediate neighbours, a property which utilises local propagation. The second uses the notion of global propagation where, in one operation, information can be passed right across the processor array. These two transform types will now be explained in more detail.

3.1 Local Propagating Transforms

The physical connections which exist between adjacent PEs allow each pixel to inform its neighbours of its present state. The neighbourhood information and the pixel's own state are combined to create the new image. There are five classes of transform within this category: edge finding, point noise removal, image shifting, shrinking and expanding.

If the image is considered to consist of white objects (sets of 1s) on a black background (0s) then the edge-finding operation can be described, in words, as: "Only white pixels with black neighbours will be white in the transformed image", and the appropriate SET instruction is

SET   A.P,[1-8]-A

It is interesting to note that the result is the 4-connected edges of the objects. This arises because only the background (0) pixels give out a propagation value of 1 and all eight neighbour inputs are enabled. The background is thus being treated as 8-connected and so any line of 1s must be 4-connected to be continuous. To obtain the 8-connected edges, the direction list would have to be modified to include only the north, east, south and west neighbours, the background then being treated as 4-connected. The problem is usually expressed as the connectivity paradox (Figure 1.5) which is only resolved if 8-connected objects exist on a 4-connected background or vice versa.

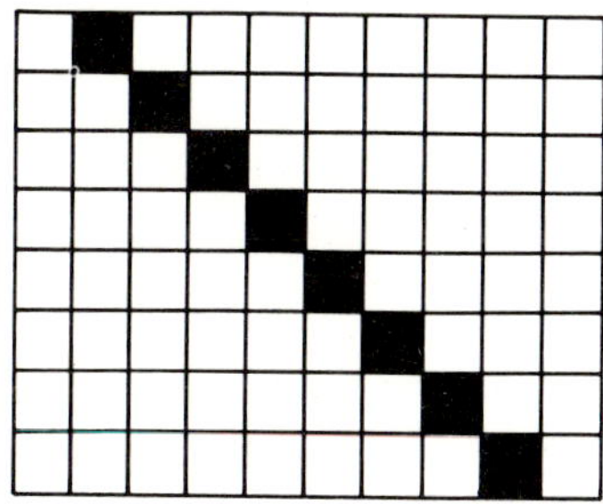

Figure 1.5   The connectivity paradox.   Does the 8-connected line divide the array into two separate regions?

Point noise can easily be removed since the noise pixels have no similar immediate neighbours. For white noise elimination the appropriate SET instruction is

SET   A.P,[1-8]A

Only the white pixels with white neighbours remain in the transformed image.

An image can be shifted along any of the PE interconnection directions using

SET   P,[d]A

where d is the desired direction of shift. This operation has been described in some detail in the previous section on the CAP4 language.

Shrinking is the process of peeling away the outer layer of pixels from an object and, conversely, expanding is adding a layer of pixels to the object. The two operations are duals of each other, shrinking the object being the same as expanding the background and vice versa. The SET instructions for shrinking and expanding are, respectively,

SET   A.-P,[1-8]-A

SET   A+P,[1-8]A

All eight neighbour directions are shown enabled, resulting in fully symmetric image transforms, but sometimes only a subset is used to give the desired result. Serra has extensively studied the shrink and expand operations, which he calls erosion and dilation respectively, and has put their use on a sound theoretical basis [13].

In all the above operations it was assumed that there were white objects on a black background. If it is desired to work with black objects on a white background then the logic of the SET instructions can be inverted, e.g. black point noise removal would be achieved using

SET   A+-P,[1-8]-A

3.2 Global Propagating Transforms

The transforms using global propagation are dependent on
the notion of there being separate connected sets within the
image.   Propagation initiated at a point or points on a
connected set passes throughout the connected set to its
borders, allowing transforms to be applied to these sets as
a whole.   It is important to note that the connected sets
existing within an image change with a change in the enabled
neighbourhood directions.   For square connectivity, if all
eight directions are used then any contiguous set of pixels
with the same value constitutes a connected set, but if,
say, only the north and south directions are enabled then
connected sets can only exist in the vertical direction;
there is no horizontal connectivity.

One of the most frequently used global propagating
operations is labelling.   Here a second image (the label
plane) is used as a source of propagation to extract desired
objects from the image to be transformed.   Again assuming
white objects (1s) on a black background (0s), any 1 pixels
in the label plane which overlap objects in the source image
initiate propagation through those objects and separate them
from any other objects.   The SET instruction for this
operation is

SET   A.P,[1-8B]A.P

The B in the direction list enables the second image register, the B-register, which holds the labels.   In words, the
process can be described as follows.   Propagation is initiated by those pixels in the B-register which have a value
of 1.   Only 1 pixels in the source image which receive a
propagation signal pass on such a signal and only the 1 pixels which receive a propagation signal remain as 1s in the
transformed image.   All other pixels become 0.

The edge of the array fulfils a special function in
several global propagating transforms.   If the E default-
option of the SET instruction is used then the PEs around
the perimeter of the array receive a propagation signal from
the edges.   Modified forms of the labelling operation can
then be used to keep or remove only objects touching the
edge of the array.   For retaining edge-connected objects
the SET instruction would be

SET   A.P,[1-8]A.P,E

while for removing them it would be

$$\text{SET} \quad \text{A.-P,[1-8]A.P,E}$$

In both cases propagation only passes through the objects touching the array edges.

If propagation is passed only by the background then two more image transforms become available. Holes in objects can be extracted since they represent parts of the background not connected to the edges. For hole finding the appropriate SET instruction is

$$\text{SET} \quad \text{-A.-P,[1-8]-A.P,E}$$

Secondly, the outer edges of objects can be found. These can be defined as those object pixels which are adjacent to background pixels which are connected to the array edges. To obtain 8-connected outer edges the required SET instruction is

$$\text{SET} \quad \text{A.P,[2468]-A.P,E}$$

Again, the background has to be treated as 4-connected to obtain 8-connected lines representing the edges. In this particular transform the desired pixels are extracted since they receive propagation even though they are not members of the connected set which is passing propagation. In the other transforms just described, the extracted pixels were members of connected sets which had passed propagation in the cases of labelling and edge-connected object extraction, or were pixels which had not passed propagation in the cases of hole finding and edge-connected object removal.

3.3 Multiple Instruction Operations on Binary Images

Most image transforms require a sequence of CLIP4 operations for their implementation. A typical example is a form of mask fitting where it is desired to find those pixels with a particular neighbourhood configuration, but where the configuration can have any orientation. Two CLIP4 operations are required to find the configurations for each possible orientation, so for a hexagonal tessellation six pairs of such operations would normally be required to find all the wanted pixels. Extra Boolean operations would then be required to combine the separate results in one image plane.

A special case of mask matching is thinning, where it is the 1 pixels that do not match the mask which remain as 1s in the transformed image. The mask is applied iteratively until a figure of one pixel width remains in the image. The mask is chosen to find those 1 pixels which can be eliminated without breaking the object into more than one connected set.   For a thinning algorithm based on a hexagonal tessellation the mask shown below can be used [14].

```
    1     1

    x  .  x            x = don't care
                            (0 or 1)
       0     0
```

This is applied in six possible orientations for one iteration of the thinning algorithm.   The required CAP4 instructions for the application of one orientation of the mask are

```
        SET   A.P,[45]A,H
        LDA   IMAGE
        PST   TEMP

        SET   A.P,[12B]-A,EH
        LDA   IMAGE
        LDB   TEMP
        PST   IMAGE
```

The first CLIP4 operation finds those 1 pixels in IMAGE which do not have 0 neighbours in the required positions and stores the result in TEMP.   The second operation is more subtle because it not only finds the 1 pixels which do not have the required 1 neighbours but includes the information about the neighbours found in the previous operation.   A more obvious implementation would be to find the 1 pixels without the required 0 and 1 neighbours in two operations and combine the results in a third Boolean operation.   An object and its thinned version are shown in Figure 1.6.

The direction list can be held in a register so that it is not necessary to write out the instructions for the application of each mask orientation explicitly.   It is worth mentioning here that the limited flexibility of the register instructions makes mask rotation a tedious process. For a hexagonal direction list the top six bits of a 16-bit register are used, a set bit corresponding to an enabled

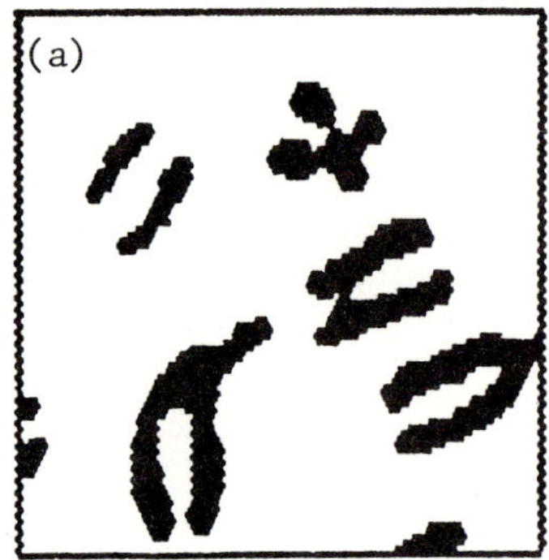

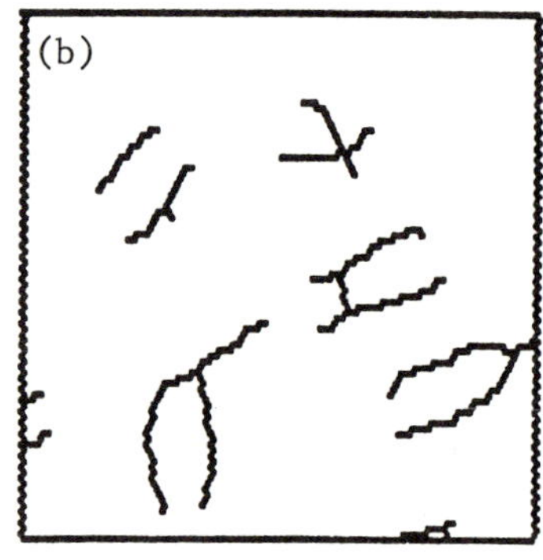

(a). Binary image of   **(b) Result of thinning.**
chromosomes.

Figure 1.6   Thinning on a hexagonal tessellation:
(a) binary image of chromosomes;   (b) result of thinning.

direction.   Bit 15 corresponds to direction 1 and bit 10 to
direction 6.   For directions 4 and 5 to be enabled a register would contain

```
15                    10                              0
 0   0   0   1   1   0   0   0    0   0   0   0   0   0   0   0
```

Rotation of a mask in an anti-clockwise manner can be
achieved by adding the register to itself, which doubles the
contents, so moving the set bits one place to the left.
However, before this is done it is necessary to test whether
the contents of the register are negative, i.e. whether bit
15 is set.   If it is set, then bit 9 can be set to 1 by
adding 1000_8 to the register and then performing the doubling.   In this way direction 1 is transferred to direction
6 during the anti-clockwise rotation.

In complete contrast to mask matching, another multiple
instruction image transform is that of finding an object's
minimum circumscribing polygon, which here will be loosely
termed its convex hull.   This process involves global propagation, rather than the local propagation used in mask
matching.   The convex hull of an object on a particular
tessellation can be found by projecting the object along the
principal directions of the tessellation.   It is the operation of projection which involves global propagation and is

achieved on CLIP4 using, for a hexagonal tessellation,

SET   A+P,[14]A+P,H

This would spread the object along the first diagonal to the edges of the array. Similar projections would be required for the other diagonal and the horizontal directions. The hexagonal convex hull is then the intersection of the projections. It will be realised that there can be only one object within the image otherwise projections from different objects would interfere. A typical result is shown in Figure 1.7.

4. ARITHMETIC ON CLIP4

Numbers can be held in the CLIP4 array in two different ways. The first is the bit-column format where each PE of a column (or row) holds one bit of the number. The array can thus hold ninety-six 96-bit numbers in one binary image plane and can process them all simultaneously. The second format is the bit-plane representation where each PE holds a complete number, each bit in a binary image plane. The CLIP4 array can process a matrix of 96 x 96 numbers held in the same manner. The numbers can be of any size, within the limits of storage, since all processing is done in a bit-serial manner. There is a direct correspondence between the bit-plane format and an image so it is this that is most commonly used for picture processing. Both formats are illustrated in Figure 1.8.

4.1 Bit-Column Arithmetic

The full-adder capability of the CLIP4 PE and global propagation allow the addition of ninety-six pairs of fixed-point bit-column numbers in one operation. The appropriate sequence of CAP4 instructions is

```
SET   -A@P,[6B]A.P,R
LDA   NUM1
LDB   NUM2
PST   SUM
```

Using direction 6 only allows propagation vertically up the array so that the least significant bits are at the bottom

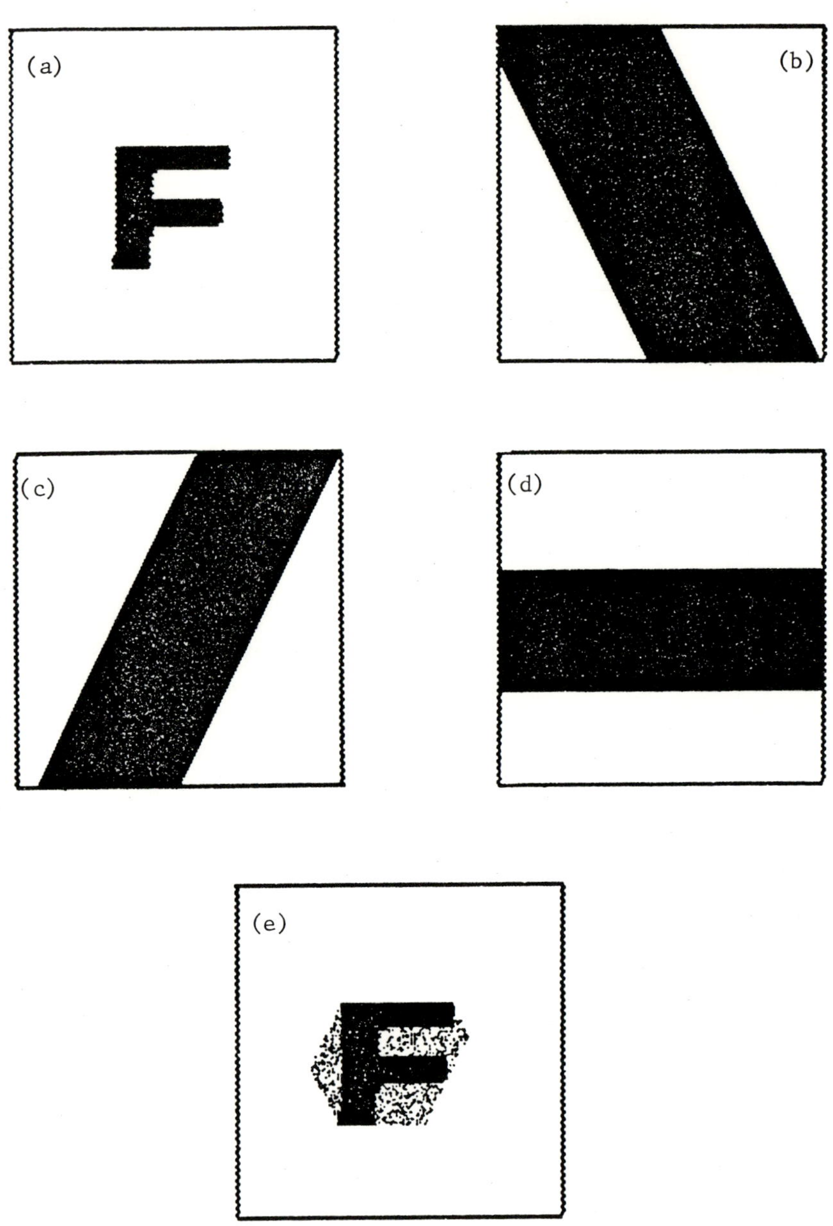

Figure 1.7   Forming the hexagonal convex hull of an object: (a) the object; (b)-(d) projections along the principal directions of the hexagonal tessellation; (e) the object and its convex hull.

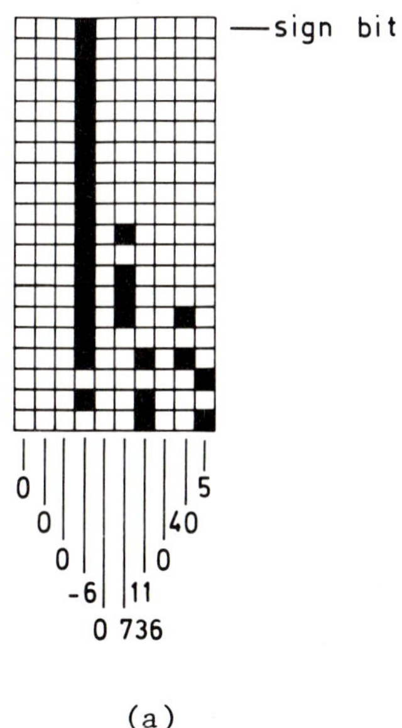

(a)

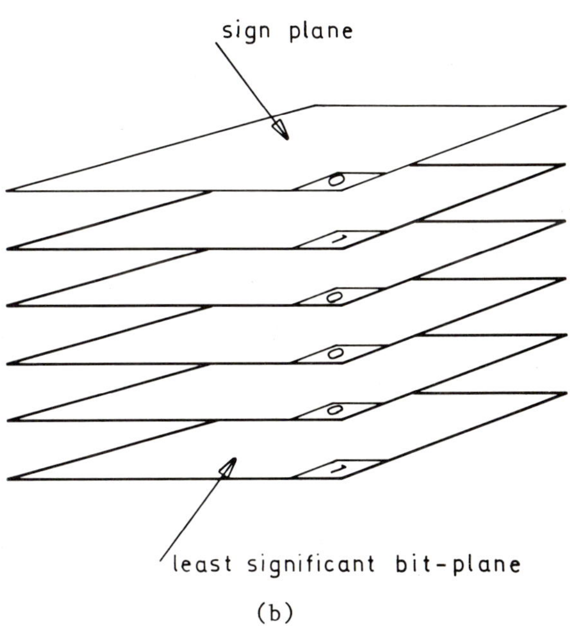

(b)

Figure 1.8   Methods of number storage: (a) bit-column integer;   (b) bit-plane integer (17 shown stored).

and the most significant are at the top. The R default-
option sets each PE as a full adder so that the P input to
the Boolean processor is the sum of the B-register and the
carry from the adjacent, less significant, PE. The D-
output definition forms the sum of the A-register and the
P-input, whilst the N-output definition produces the carry
to the next, more significant, PE.

Subtraction can also be performed in one operation.
Two's-complement arithmetic is used but it is not necessary
to form the complement in one operation and then perform an
addition in a second operation. The two's complement of a
number is usually formed by logically inverting each bit of
the number and then adding 1. By choosing the appropriate
D and N functions the inversion is done by the Boolean processor and the 1 to be added is provided by setting the
edges of the array to 1. The minuend must be placed in the
B-register and the subtrahend in the A-register, the
required sequence of CAP4 instructions being

```
SET  A@P,[6B]-A.P,RE
LDA  SUBEND
LDB  MINEND
PST  DIFF
```

The time required to execute either an addition or subtraction in CLIP4A is completely dominated by the unnecessarily
long global propagation time, and is approximately 630 μs in
each case.

Multiplication of bit-column numbers is performed in the
usual manner of repeated shifts and additions except that,
on a given addition operation, generally only some of the
multiplicands will have to be added to the partial products
plane. To do this the appropriate multiplicands must be
selected and placed in a temporary storage plane, the
columns corresponding to the multiplicands not required for
this addition being left as zero. The selected multiplicands are then added to the partial products plane. A flow
chart of the required operations is shown in Figure 1.9.
All numbers have a fixed length of 96 bits and so the product must never exceed 96 bits or errors will occur. Since
the product is the same fixed length as the multiplicand and
multiplier then multiplication of negative numbers held in
two's complement can be performed directly. If only one of
the operands is negative then multiplying the two gives a
negative product in two's-complement form:

$$x(2^{96} - y) = x2^{96} - xy = (x - 1)2^{96} + (2^{96} - xy)$$

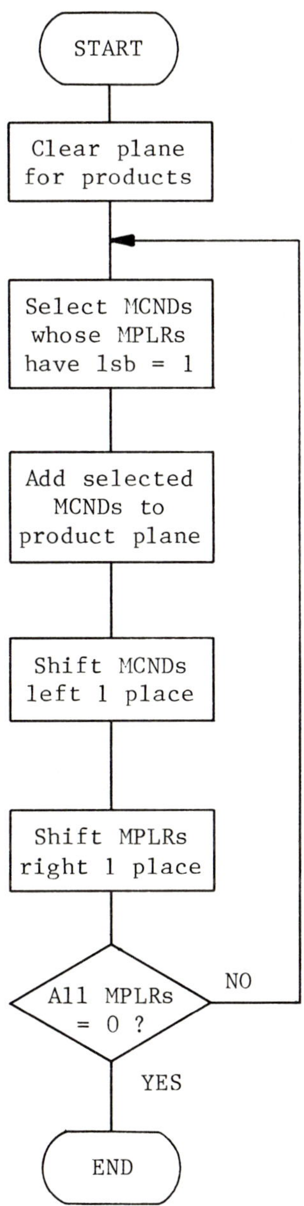

Figure 1.9   Bit-column multiplication (MCND = multiplicand, MPLR = multiplier, lsb = least significant bit).

where y is the negative number. The $(x - 1)2^{96}$ term over-
flows the column and is lost. Similarly, if both operands
are negative, then multiplying yields a positive product:

$$(2^{96} - x)(2^{96} - y) = 2^{192} - 2^{96}(x + y) + xy$$

All terms except the product can be ignored since they overflow and are lost.

The 96-bit length of numbers allows accurate answers
even for fixed-point division which has been implemented on
CLIP4 using the non-restoring division technique described
in [15]. A flow chart of the basic process is shown in
Figure 1.10. The dividend is shifted left into a special
accumulator and, after each shift, the divisor is subtracted
from or added to the accumulator according to the rules of
non-restoring division. As the dividend is shifted out the
quotient is formed in the bottom of the dividend register.
If the last accumulator operation gave a negative answer
then a 0 is inserted in the quotient, otherwise a 1 is
inserted. The dividend is shifted left until it has all
entered the accumulator, which would take ninety-six iterations on the CLIP4 array.

Dealing with ninety-six numbers simultaneously adds to
the problem of implementing the division algorithm since, on
any given iteration, some divisors will have to be subtracted and the rest added, and 1s will need to be inserted
in only some of the quotients. The solution to the first
problem is to negate the divisors which are to be subtracted, leaving the rest unchanged. This modified set of
divisors is then added to the accumulator plane, giving
addition and subtraction in the appropriate places.
Inserting the correct values into the quotients is achieved
simply by using the logical inverse of the sign bits resulting from the last accumulator operation. Division of negative numbers is catered for by forming the absolute values
of the divisors and dividends before division begins. The
correct signs are given to the quotients when division is
complete. The flow chart for the full algorithm is shown
in Figure 1.11. If extra places of precision are required
in the quotients then the dividends are left-shifted the
necessary number of places before division begins, e.g. if
the divisor and dividend planes held intégers but the quotients were required to have twenty fractional binary places
then the dividends would have to be left-shifted twenty
places. The division is still performed through ninety-six
iterations.

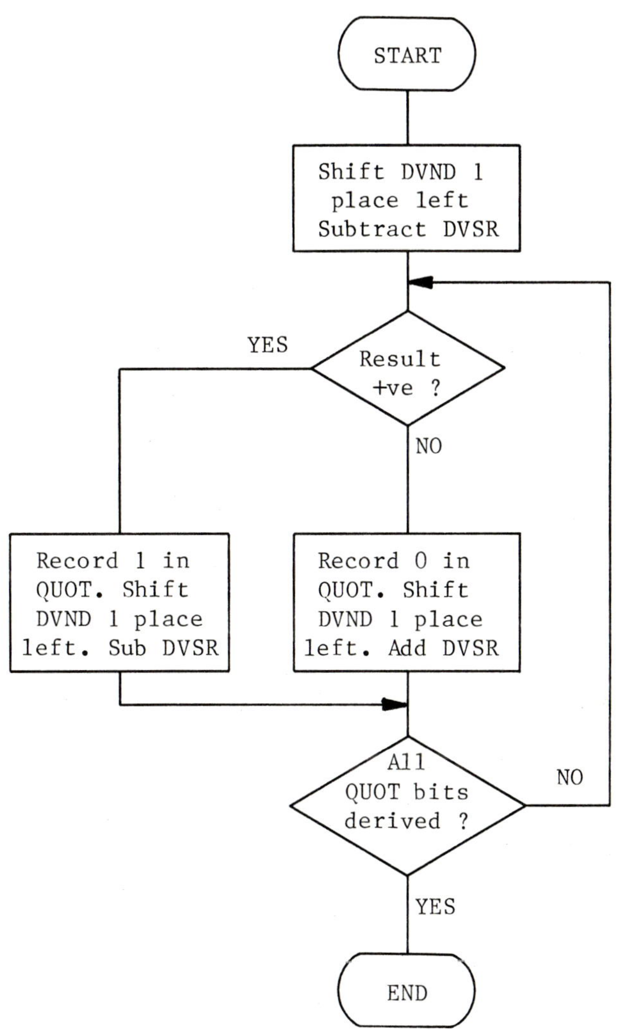

Figure 1.10 Non-restoring division flow chart (DVND = dividend, DVSR = divisor, QUOT = quotient).

                    S. D. PASS

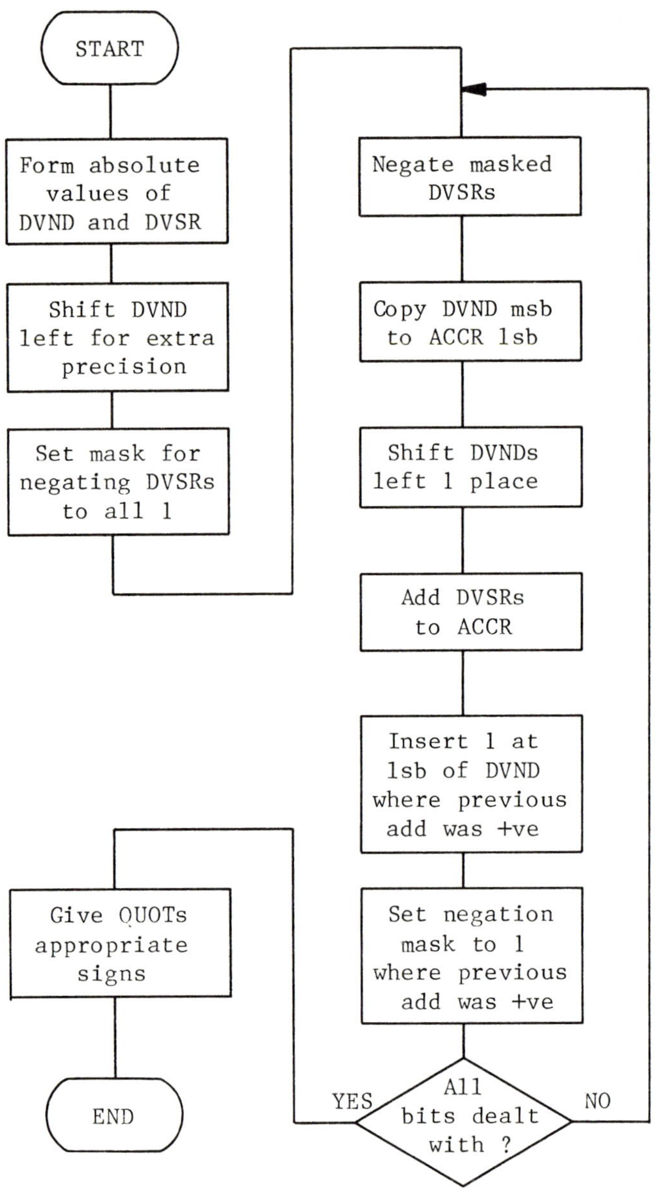

Figure 1.11   Bit-column non-restoring division (ACCR = accumulator, DVND = dividend, DVSR = divisor, QUOT = quotient, lsb = least significant bit, msb = most significant bit).

The algorithms used for forming the absolute value of all the numbers in a plane and for negating selected numbers are worthy of explanation. They are very similar since forming the absolute value is no more than selective negation, the negation mask being obtained from the sign bits of the numbers. For this reason only the selective negation algorithm will be described. Ones are placed in the negation mask in the columns that are to be negated but, for the algorithm, it is the columns that are not to be negated which require a 1 in the mask and those to be negated a 0. This is achieved by logical inversion. The heart of the process is the following SET instruction (the direction is arbitrary)

SET   A@P,[6BC]A+P,R

The A-register is loaded with the plane to be selectively negated and the B-register with the inverted negation mask. The whole column corresponding to a number to be left unchanged must be filled with 1s, the other columns with all 0s. Before this SET instruction is executed the C-register must also be loaded with columns of the inverted mask values. The operation of the SET instruction is best understood with reference to the CLIP4 PE diagram (Figure 1.1) and the 5-bit examples which follow. As the R default-option is used, the gates producing the P input to the Boolean processor are set as a half-adder. The P-input is then the exclusive OR of the B-register and the neighbourhood input. The values loaded into the B- and C-registers are such that only the selected numbers are negated.

Copy number		Negate number	
A-register	0 0 0 1 0	A-register	0 0 0 1 0
B-register	1 1 1 1 1	B-register	0 0 0 0 0
C-register	1 1 1 1 1	C-register	0 0 0 0 0
P input	0 0 0 0 0	P input	1 1 1 0 0
N output	0 0 0 1 0	N output	1 1 1 1 0
N* output	1 1 1 1 1	N* output	1 1 1 1 0
D output	0 0 0 1 0	D output	1 1 1 1 0

Input = +2   Input = +2

Result = +2   Result = −2

From division it was a simple matter to implement a square root algorithm using the Newton-Raphson iterative procedure. The value of the nth approximation to the square root is given by

$$x_n = \frac{1}{2}\left[\frac{y}{x_{n-1}} + x_{n-1}\right] \rightarrow \sqrt{y}$$

The procedure is performed until only the least significant bit of the square root changes between successive iterations. This condition is necessary because, when using fixed-length arithmetic, convergence to a static value is unusual, the least significant bit often oscillating between 1 and 0.

The most innovative part of the algorithm is the method of obtaining the first approximations to the square roots. Right-shifting the number to be rooted (division by 2) until it is half its original length provides a reasonable first approximation. However, for a plane of binary column numbers, each number will generally have to be shifted a different amount. The shift-counting is achieved by taking the numbers to be rooted and filling the columns from the most significant bit of each number downwards to create a block histogram of number lengths (Figure 1.12). Bi-directional shrinking is applied to the histogram to remove the top and bottom of each column in one step and for each shrink operation, except the first, the numbers to be rooted are right-shifted one place. To prevent numbers of only one bit becoming 0 (which would lead to division by 0), the shifting must be performed once less than the number of shrinking operations. When a histogram column becomes zero the number it corresponds to is no longer shifted, the number of steps taken to eliminate a column being equal to half the column length.

The efficacy of this first approximation method can be shown by examining the number of iterations taken to reach convergence. Numbers from 1 to 17 were loaded into the columns of a plane. It was initially required to calculate roots to 10 fractional binary places. Using the number itself as the first approximation took 21 iterations to reach convergence, but using the half-length number took only 4 iterations. The roots were then calculated to 40 fractional binary places and the corresponding numbers of iterations were 46 and only 5. The half-length approximation thus speeded the algorithm considerably. The value

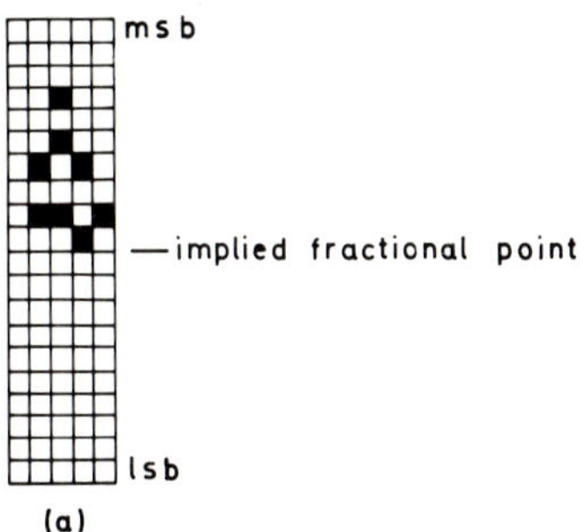

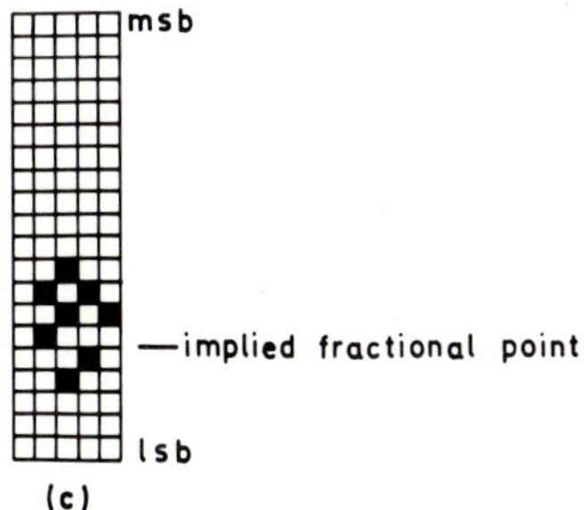

Figure 1.12   Finding the first approximation to the square root: (a) the numbers to be rooted (10, 82, 9, 2 from the left); (b) number-length histogram; (c) first approximations (5, 10, 25, 4.5, 2 from the left).

obtained for $\sqrt{2}$ to 40 binary places was

$$1.011\ 010\ 100\ 000\ 100\ 111\ 100\ 110\ 011\ 001\ 111\ 111\ 001\ 1$$

which is approximately 1.4142135628.

4.2 Bit-Plane Arithmetic

CLIP4 performs arithmetic on numbers held in the bit-plane format in a bit-serial manner. A pair of bit-planes is added (or subtracted) in one operation. The instructions required for addition and subtraction are very similar to those for bit-column arithmetic except that they have to be repeated to process all the bits in the numbers. The carry resulting from the addition of a pair of planes is held in the C-register for addition with the next pair of planes. The SET instruction required for addition is

```
SET   A@P,[BC]A.P,R
```

The C in the direction list allows the contents of the C-register to pass through the neighbourhood input gating and effectively takes the place of the direction number in the column addition SET instruction which allowed the carry to propagate from one PE to the next.

Subtraction is performed in two's-complement arithmetic but, as for column subtraction, the complement of the sub-trahend can be formed as the subtraction is performed. The C-registers of all PEs must be initially set to 1 to provide the 1 to be added to the logically inverted subtrahend. This can be done with the following instructions

```
SET   A,1,R
LDA   IMAGE
PST   IMAGE
```

A PST instruction must be performed to load 1 into the C-register but to avoid corrupting any D-memory planes the A-register is loaded with an arbitrary image. The SET instruction is so defined that when the PST is executed the image is left unchanged and is stored in the location from which it came. To perform the subtraction the appropriate SET instruction is

```
SET   -A@P,[BC]-A.P,R
```

Addition and subtraction take approximately 34 μs for each pair of operand planes, plus a little more for initially setting up addresses and loop counters.

It will be realised that the 32 planes of storage which CLIP4 possesses are soon exhausted when performing bit-plane arithmetic. Fixed-point multiplication and division have been implemented using algorithms similar to those for bit-column arithmetic described previously but they can only be used to low precision. However, proposals for special back-up storage to be implemented on later CLIP machines should considerably enhance the arithmetic capability of the CLIP family.

CHAPTER TWO

PROPAGATION   IN   CELLULAR   ARRAYS

A special feature which distinguishes the CLIP4 array from others of its kind is a provision for the global propagation of data through the array. The operations which permit this fast passage of data through distances of many pixels help to overcome one of the fundamental limitations inherent in locally connected networks of processors. They are therefore of considerable importance but are. unfortunately, amongst the most difficult functions to understand.

A substantial part of Paul Otto's thesis 'Algorithms for Image Processing on the CLIP4 Cellular Array Processor' is concerned with the understanding and use of these global propagation operations. The material in this Chapter is drawn from various sections of his thesis and presents a coherent description of two important functions of this type namely object counting and object labelling.

1. INTRODUCTION

Global array instructions in CLIP4 are those in which
the data sent to neighbouring cells is a function of the
data received from neighbouring cells and from the point
itself.   They permit information to be passed over arbitrarily large distances in the array in one instruction, so
that the result is potentially a function of the data in the
whole array.

1.1 An Example of Global Propagation

Figure 2.1 shows some letters: let us assume that we
wish to delete all letters which do not have loops.   The
background can be divided into the area which is connected
to the edge of the array, and the areas which are not (i.e.
the holes in the letters).   Clearly the letters which have
loops are exactly those letters which touch the holes, so
the first step is to identify the holes.

Figure 2.1   Some letters.

If we set the edges of the array to 1 (so that the propagation input to each cell at the edge of the array is 1), and output a 1 from each cell if it receives a 1 and is part of the background, then we will send a signal from the edge of the array, through all background points which are connected to the edge, up to the edges of the letters. This means that any background point which does not receive a signal cannot be connected to the edge, i.e. it is part of a hole.

More formally, at each cell, we simultaneously execute the following two operations:

$$\text{propag\_in} \; . \; \overline{\text{letter}} \rightarrow \text{propag\_out}$$

$$\overline{\text{propag\_in}} \; . \; \overline{\text{letter}} \rightarrow \text{hole}$$

where propag_in is the (binary) value propagated into the cell, propag_out the value propagated out, . denotes the Boolean AND function, and so on. This gives us the holes (Figure 2.2(a)) in one instruction.

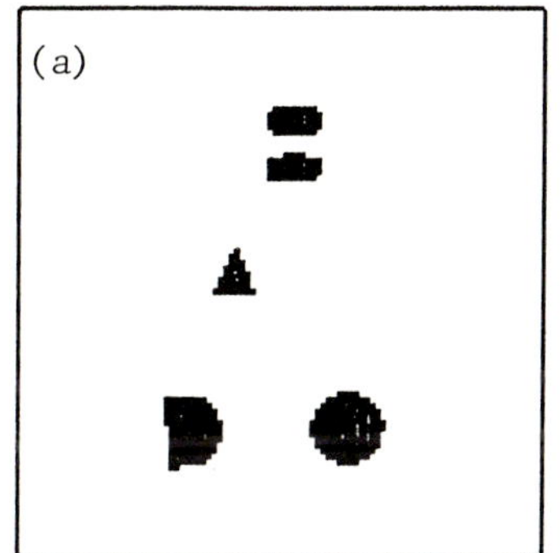

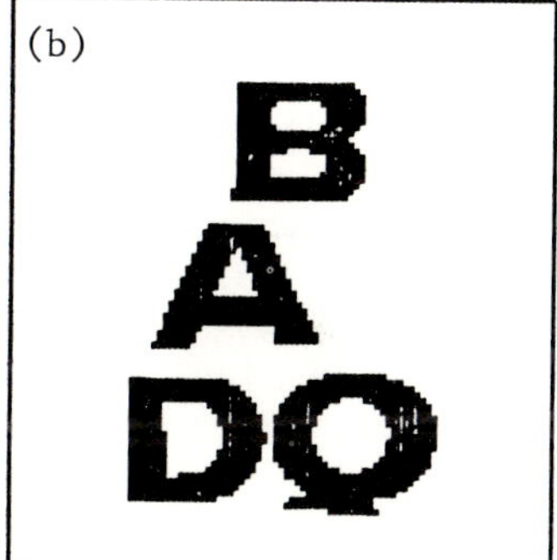

Figure 2.2   (a) The holes in the letters in Figure 2.1. (b) The letters in Figure 2.1 which have loops.

Now we have to pick out the letters which touch the holes. This time we send out a signal from all points which are part of a hole, allowing it to propagate through points which are part of a letter. Then all points which receive a signal, and are part of a letter, must be part of a letter which touches a hole. In other words these points are the points of the letters which have loops, which was the

required result (Figure 2.2(b)). In formal terms, this step
is:

$$\text{hole} + \text{propag\_in} \; . \; \text{letter} \rightarrow \text{propag\_out}$$

$$\text{propag\_in} \; . \; \text{letter} \rightarrow \text{loopyletter}$$

One subtlety to note about these operations is that if
we wish (for example) to regard the letters as 8-connected,
and the background as 4-connected, then we must enable only
the horizontal and vertical directions (2468) when propagating through the background, and enable all directions (1-8)
when propagating through the letters.

2. COMPUTATIONAL COMPLEXITY ON CLIP4

We wish to be able to get some estimate of how good an
algorithm is. The most interesting criteria are usually how
much time or space an algorithm requires for given input
data.

The time (for example) required by an algorithm is frequently expressed in two different ways: as the time
required for some particular input, and as the way in which
the time varies as a function of the size of the input data
[16]. Both are of interest to us. We are also interested
in how the time varies as a function of the array size, so
when discussing the algorithms we will use X to denote the
width of the array, and Y for the height (i.e. the array
will be X x Y).

Clearly the complexity of an algorithm executed on CLIP4
is rather different from the complexity when executed on a
serial machine, so we will simply use the CLIP4 instruction
count as our measure of the time required for an algorithm.
However, we will not count the number of serial instructions
for two reasons: the time used by the serial instructions
(in the algorithms we are going to consider) reflects the
structure of the controller rather than the algorithm; and
the proportion of the total time used by the serial instructions is usually small.

Moreover, we will tend to concentrate on the number of
global operations required, rather than the total number of
array instructions. This is partly because a typical global
operation takes longer than a local or pointwise operation,
and partly because the number of pointwise operations

required depends on details of the CLIP4 cell, whereas the number of global operations does not.   (The number of global operations depends primarily on the interconnections between cells.)   Note however that, in these algorithms, the time required for the non-global operations is comparable to or much less than the time used by the global instructions, simply because a large proportion of the operations are global.

Approximate times (in μs) for the various instructions using the prototype CLIP4 chips are:

Pointwise array operation	9-34
Local array operation	9-34
Global array operation	630
Count operation	110
Serial instruction	2-6

Global propagation instructions should only take about 9-34 μs + 3 μs for propagation through each cell, but they take longer because of faults in the prototype array chips. These faults interfere with the mechanism for detecting the completion of propagation, as a consequence of which a timeout is used in the prototype system and is set at 630 μs to allow for a sensible longest probable propagation path. The production (CLIP4D) chips generally run about 2.5 times faster than the prototypes, and work correctly for global operations, so that a pointwise or local operation takes about 4-12 μs, and a global operation takes about 4-12 μs + 0.2 μs per pixel.

A global operation can be regarded as a repeated local operation;   however, a global operation is much faster because it does not have the repeated overheads of instruction fetching/decoding and data fetching/storing.   This means that an iterated local operation can only propagate information at about one cell per memory cycle, whereas a global operation is only limited by gate switching times, so that it can easily be an order of magnitude faster than an iterated local.   For the CLIP4D chips, a global operation propagates about 50 times faster than a repeated local, and so one which sends a signal across the whole array only takes about three times as long as a single local operation. This justifies the treatment of all array operations as being of comparable complexity.

3. COUNTING OBJECTS

Throughout this Chapter we shall use `object' to denote a connected non-zero region of a binary image. We will also be using 8-connectivity for objects and 4-connectivity for background unless otherwise stated.

The immediate problem we pose is to discover how many objects there are in a given binary image. Before we start to consider solutions to this problem, an example will give us a better feel for the task.

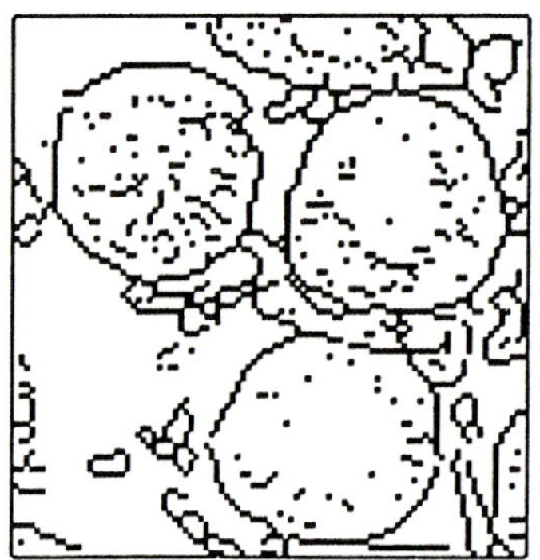 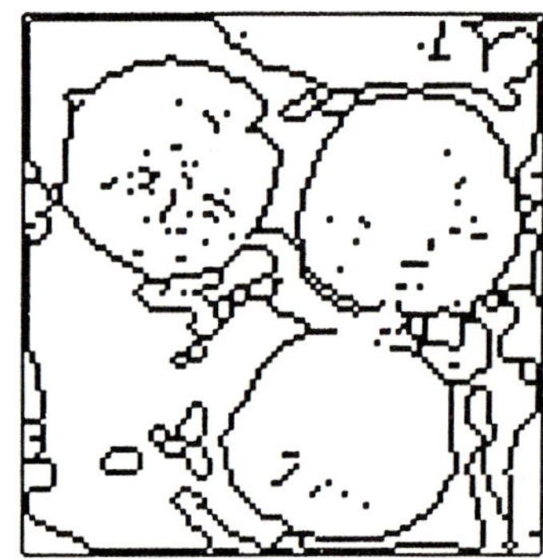

Figure 2.3   Partially segmented cell images.

The pictures in Figure 2.3 both show an early stage in segmenting cell images, using a semi-automatic segmentation program developed by Reynolds [17] and described in Chapter 8. Given a set of images and some fairly simple guidance by the user, this program tries to produce a segmentation algorithm/program for that class of images. In order to guide its attempts at segmentation, the program often needs to measure several parameters of the partially segmented image, among which are the number of regions of various sizes, and the number of line segments. Note that in this case the objects of interest (8-connected black lines and 4-connected white regions) are not well separated compact blobs, but intertwined regions which sometimes extend over large areas of the image, or surround one another. Two promising approaches to the counting problem are:

1. Select an object in the image, increment a counter, delete the object and repeat until finished.

2. Reduce objects to single points and then count these points.

Approach (2) looks better than approach (1) for two reasons: it looks more parallel so that it should make more efficient use of the machine; and its execution time should be more or less independent of the number of objects, whereas the time taken by (1) will clearly increase linearly with the number of objects. However, it is not quite as simple to reduce objects to single points using a cellular array processor as it appears at first, especially if the objects have convoluted shapes or are nested within each other as in Figure 2.3. On the other hand, each iteration of approach (1) is very simple, so we will describe that first.

3.1 Counting by Sequential Removal

Algorithm count_obj_1

1. Set count (an integer variable) to zero.

2. If **image** is zero everywhere, have finished.

3. Find a point in an object in **image** using the top_left_point algorithm described below.

4. Increment count.

5. Delete object connected to selected point in **image.** (Send a propagation signal in all directions from the selected point, transmitting the propagation signal only at points where **image** is non-zero, and then delete any point of **image** which receives a signal.)

6. Go to (2).

The basic idea for selecting a point in the object is very simple: each point sends a signal downwards and to its right; then the point which does not receive a signal is the topmost leftmost point. It is only slightly more complicated in practice.

Algorithm top_left_point

1. Send propagation signals from all non-zero points of **image,** and receive them in directions 1-3 and 8, keeping only those points which do not receive a propagation signal. (All points which receive a signal pass it on.)

2. Send propagation signals from all surviving points, and receive them in directions 1-4, keeping only the point which does not receive a signal. (That is, repeat step (1), using different directions, on the result of step (1).)

Clearly there will be at least one point in the result (unless **image** was zero everywhere initially), because a point is only deleted if there was another point above or to the left (or, in step (2), to the right). On the other hand, there will be at most one point in the result, since any surviving point will have deleted all points at the same level or below, and would have been deleted itself if there were any points above it. Thus, in two steps (each a single global propagation instruction), this algorithm selects a single point in a binary image. Figure 2.4 shows the various steps of top_left_point on a simple image.

Thus count_obj_1 only needs three global propagation instructions per object. This is quite fast if the number of objects is small, but depressingly slow if the number of objects is large.

3.2 Counting by Reduction to Points

Let us go back and consider approach (2) to the counting problem, i.e. reducing objects to points and then counting the points. The difficulty here is to find an efficient way of reducing objects to points; counting the points is easy, especially as there is built-in hardware to do just this. If all the objects are simply connected, i.e. are objects without holes, then we can just shrink the objects, preserving connectivity, until they become points. For example, in hexagonal connectivity, we can delete all points at which the mask

```
  0 *
0 1 1
  0 *
```

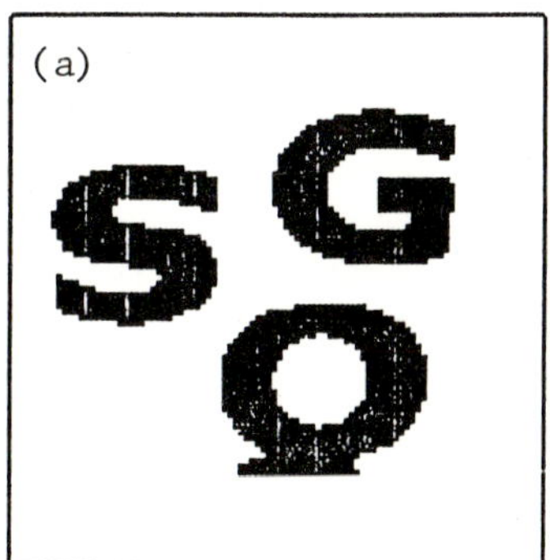

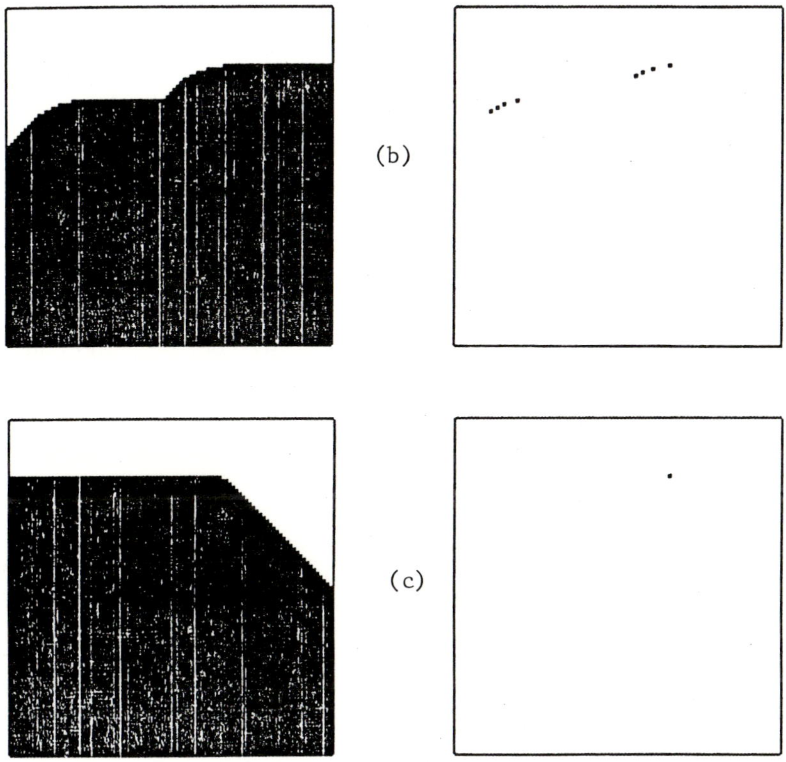

Figure 2.4   (a) Some letters.   (b) The propagation signal, and the remaining points, after step (1) of the top_left_point algorithm.   (c) The propagation signal in step (2), and the final result, when applying the top_left_point algorithm to (a).

fits (* denotes don't care), repeat this for all rotations of the mask, and then repeat the whole cycle until no more points can be deleted.   With a little more thought, parallel shrinking algorithms capable of handling any shape of object can be devised, such as those described in [18] or [19]. However, such algorithms require at least O(length of object) operations and cannot easily be extended to handle arbitrarily nested objects.

Any algorithm for collapsing objects to points will require information from arbitrarily large regions of the array (depending on the input data), so it seems likely that using global propagation could greatly reduce the number of operations required. This thought prompts us to consider ways of generating the required points using global propagation.   Unfortunately, we cannot just generalise the top_left_point algorithm by restricting the propagations to lie within objects, since this would only work correctly with convex objects.

We have two ways of controlling which points a propagation signal goes to: we can allow propagation only at marked points, or we can enable propagation only in certain directions. In order to distinguish between points in an object (so that we can select one), we need to use some global operations which only propagate in a few directions; but if we also restrict these propagations to lie within the object, then we cannot ensure that the signals will reach all points of the object in a fixed number of operations. (Consider an object which is a large spiral, for example.)

On the other hand, restricting global propagation to lie within objects is much the fastest way of checking whether two points are part of the same object. This leads us to consider methods which use both types of global propagation instruction, such as: (1) assign a unique label to each point of the array (such as some function of its x,y coordinates), and then (2) find the point within each object which has the largest label. Clearly this will select exactly one point in each object.   We will now consider how to implement it on CLIP4.

Labelling each point of the array

First, let us generate the x co-ordinate at each point
of an 8 x 4 array:

```
0 1 2 3 4 5 6 7
0 1 2 3 4 5 6 7
0 1 2 3 4 5 6 7
0 1 2 3 4 5 6 7
```

The least significant bit-plane of this is

```
0 1 0 1 0 1 0 1
0 1 0 1 0 1 0 1
0 1 0 1 0 1 0 1
0 1 0 1 0 1 0 1
```

which is easy to generate by sending a propagation signal
from left to right in the array, starting at 0, and flipping
between 0 and 1 at each point. That is, at each point, the
output propagation signal is 1 if the input is 0, and 0 if
the input is 1.

The next bit-plane we require is

```
0 0 1 1 0 0 1 1
0 0 1 1 0 0 1 1
0 0 1 1 0 0 1 1
0 0 1 1 0 0 1 1
```

This is obtained by marking where the transition from 1 to 0
occurs in the previous (least significant) bit-plane as we
move from left to right, which gives us

```
0 0 1 0 1 0 1 0
0 0 1 0 1 0 1 0
0 0 1 0 1 0 1 0
0 0 1 0 1 0 1 0
```

A signal is then sent from left to right, flipping at the
non-zero points. At each point, the output signal is the
exclusive OR of the input signal and the marker.

The next plane is

```
0 0 0 0 1 1 1 1
0 0 0 0 1 1 1 1
0 0 0 0 1 1 1 1
0 0 0 0 1 1 1 1
```

which we generate in the same fashion as the previous plane, by noting the transitions from 1 to 0, and then sending a propagation signal which flips at the marked points. By repeating this, we can generate as many planes as we need for the x co-ordinate; this number is usually approximated by log X.

The y co-ordinates are generated in the same way, propagating up the array instead of across it. A suitable function of x and y for the labels is

$$\text{label} = x + y * 2^{\left(\left\lfloor \log_2(X-1) \right\rfloor + 1\right)}$$

Computation of this label does not require any further operations since it is just a concatenation of the bit-stacks representing the x and y co-ordinates. To label a point in this way requires one global and one local operation per plane of label, and we need 14 planes for the 96 x 96 CLIP4 array (being log X + log Y).

Finding the point in each object with the largest label

To find this point, we compare the label values at different points in each object by working down the bits of the labels. If the bits are not equal, we discard the points whose label bit is 0. To illustrate this procedure, let us find the largest of the following five binary numbers:

$$0111 \quad 1101 \quad 0011 \quad 1001 \quad 1100$$

We start by comparing their most significant bits (msb):

$$0 \quad\quad 1 \quad\quad 0 \quad\quad 1 \quad\quad 1$$

As some of the numbers have a 1 as their msb, numbers with 0 as their msb cannot be the maximum so we discard them. We are left with

$$1101 \quad 1001 \quad 1100$$

We now compare their next msbs and again discard any number which has a 0 bit leaving

$$1101 \quad 1100$$

This process continues until we are left with the largest number:

$$1101$$

On CLIP4, we use a binary plane to mark the points which are currently candidates. At the beginning, all the points in the objects are possible candidates, so we set the marker plane equal to the original image.   Then we start with the most significant plane of the label and send a propagation signal from all points which are eligible and have 1s as the msbs of their labels. We only allow the signal to propagate through points where the original image is 1, so that a point in an object will receive a signal if, and only if, there is at least one point in that object whose msb is 1. Any point having a msb of 0 and receiving a signal cannot be the maximum within the object, so we delete it from the marker plane; all other points are still eligible candidates, since their msb is the same as the msb of the maximum.

Now consider the next most significant plane of the label and send a propagation signal from all eligible points for which this bit of the label is 1. Again we can delete from the marker plane the points which are 0 and receive a signal. Repeat this until we have used the least significant bit of the label.

This procedure will delete points from the marker if, and only if, their labels were less than that of another point in the same object. The resulting points will be the points at which the maximum label in each object is to be found. As all the points had different labels, this must select exactly one point in each object, as required. A by-product of this algorithm is that it propagates the value of the maximum to each point of the object, so that we could have labelled the objects at the same time.

Overall, this method of generating points uses two global, one local, and two pointwise operations per plane of intermediate label, and requires $\log X + \log Y$ planes of intermediate label; therefore the number of operations is independent of the number (or complexity) of the objects to be counted, and of their nesting. For large numbers of objects, this is much better than the counting by sequential removal algorithm, and we can speed it up further by using less wasteful labels (e.g. just labelling likely points). However, we will leave these refinements until the section on labelling since they will arise more naturally there, and because there is another, completely different, algorithm for counting objects which is faster in almost all situations.

3.3 Counting by Using the Connectivity Number

We define the connectivity number of an image as the number of objects minus the number of holes in objects. It turns out that this connectivity number (sometimes called the Euler number, or genus, or Euler-Poincaré constant) can be calculated very easily on CLIP4, because

$$\text{connectivity number} = N\begin{bmatrix} 1 & 0 \\ 0 & 0 \end{bmatrix} - N\begin{bmatrix} * & 1 \\ 1 & 0 \end{bmatrix}$$

where $N[\#]$ is the number of points where mask $\#$ fits. The masks differ for different connectivities; see [20, 21, 22]. The proof can be found in [22].

Using this formula we can calculate the connectivity number on CLIP4 using only two local operations and two count instructions. If the objects do not have any holes, then the connectivity number equals the number of objects, so this equation gives the number of objects directly, and very quickly (300 µs using CLIP4).

We can remove the holes in objects, using the global propagation instruction, by only keeping the background where it is connected to the edge, i.e. by sending a propagation signal from the edge, through points which are zero, and filling in all points which do not receive a signal. However, if an object is nested within another, then it will be joined to the surrounding object by this procedure. This does not always matter, since the included regions are sometimes considered to be part of the including object. However, in general, we will have to count the outermost objects first, then remove them (using the fact that they are the objects in the original image which are adjacent to the background in the filled image), and then repeat the procedure. This leads to the following algorithm:

Algorithm count_obj_2

1. Set count to zero.

2. If **image** is zero everywhere, have finished.

3. Fill in the holes in **image**.

4. Get the connectivity number of the filled image and add it to count.

5. Delete the outermost objects in **image**. (Send a propagation signal from the background (zero points) of the filled image, through the non-zero points of **image**, deleting all the points which receive a signal.)

6. Go to (2).

Even in its full generality, this is simple and fast, requiring only two local, two global, and two count operations per iteration. Moreover the number of iterations needed is usually small, even when the number of objects is large.

Note that the time required for an iteration of this algorithm is about the same as that required for an iteration of the sequential removal algorithm, but that the latter requires more iterations. This is because it only removes one object per iteration and therefore sequential removal is never faster than the operation based on connectivity number and is usually much slower.

The connectivity number algorithm is often faster than reduction to points; however, for the worst case, the connectivity number algorithm will require $O(X)$ iterations, the maximum number of nested objects on an X x Y array being $\min(X,Y)/4$. The reduction to points method always requires $O(\log X)$ operations, so it gives a better upper bound on the number of operations required.

4. LABELLING OBJECTS

We wish to assign distinct labels to each point of each object in an image. The simplest approach is to label the objects one at a time - as with the sequential removal counting algorithm, but assigning a label to the selected object at each iteration.

A more parallel algorithm faces the problem of ensuring that the labels are distinct. As we saw in section 3.2, we can solve this problem in a crude fashion by assigning labels based upon the position of the object in the array. In that section, we labelled points using their x and y

co-ordinates, so that each point had a distinct label, and
then selected the label of one of the points in the object
to label the object. This gives a set of object labels
which are likely to be widely separated (rather than running
1,2,3,...,N). The main drawback to this is that it gives
labels which are much larger than necessary, which costs
time and memory in later processing.

4.1 A Fast Labelling Algorithm

We can reduce the size of the labels if we do not try to
give distinct labels to all the points of the array. This
gives the following strategy:

1. Select at least one point in each object (but
    the fewer the better).

2. Assign distinct labels to these points.

3. Use the label on one of the points in the object
    to label the object.

Selecting the points

A quick way of selecting points is to use the points at
which a suitably chosen mask fits. For example,

```
0 0 *
0 1 *
0 * *
```

must fit any object in at least one place, but will only fit
at a few points. As we discussed in section 3.2, there are
many other methods of reducing objects to points, but they
are slower, and will seldom reduce the number of points
much, so we will use this mask as a first attempt.

Labelling the points

Now we must assign distinct labels to a relatively small
number of isolated points. Ideally, the labels should be
consecutive (i.e. 1,2,3,...,N for N points). If the array
were one-dimensional, this would not be difficult; we could
adapt the routine for generating x co-ordinates so that the
least significant bit changes only at the selected points.
For example, if the selected points are

```
0 0 0 1 0 0 1 1 1 0 1 1
```

then we wish to get

$$0\ 0\ 0\ 1\ 0\ 0\ 2\ 3\ 4\ 0\ 5\ 6$$

but since the value only matters at the marked points,

$$0\ 0\ 0\ 1\ 1\ 1\ 2\ 3\ 4\ 4\ 5\ 6$$

is equivalent.

The least significant plane of this is

$$0\ 0\ 0\ 1\ 1\ 1\ 0\ 1\ 0\ 0\ 1\ 0$$

which is obtained from the original by propagating from left to right, flipping between 0 and 1 at the marked points. The more significant planes can then be derived from this in the same way as the more significant planes of the x co-ordinate were derived from the least significant plane.

This process can be extended to two dimensions and we can use it to distinguish points in (say) the same row. We then need to distinguish points in one row from points in another; in other words, we need to label the rows. An analogy, which may help the reader, is to think of the marked points as houses and the rows as streets. The row labels are then the street names and the label of each point in a row is the number of the house, so that a full address is given by specifying both the street name and the house number.

A neat way of labelling the rows would be to label each row with the number of marked points in the rows beneath it. We could then combine these labels (the row label, and the label of the point within each row) to get a single label for each point just by adding them together. This single label would be equal to the number of marked points below and to the left of the given point. (In terms of the analogy, we would use the street name and house number to assign a postcode to each house, which would completely specify the address just by itself.) Note that the resulting labels would be consecutive.

We wish the label on the jth row to be L given by

$$\sum_{i=0}^{j-1} n(i)$$

where $n(i)$ denotes the number of points in the ith row.

We can calculate n(i) using the routine above, while we are labelling the points within each row, the largest label in each row being n(i).

To obtain L we need to add together two numbers at each point: the number originally at the point n(i) and the partial sum which is being propagated in $\sum_{i=0}^{j-1} n(i)$. We then propagate this sum to the next point, and store the value of the partial sum. We could store either the incoming or outgoing partial sum (i.e. $\sum_{i=0}^{j-1} n(i)$ or $\sum_{i=0}^{j} n(i)$), but the outgoing sum is marginally more useful in general, so we will give the variant of the algorithm which stores the outgoing sum. (Note that we can get $\sum_{i=0}^{j-1} n(i)$ from $\sum_{i=0}^{j} n(i)$ just by doing a shift, but not vice versa, because the last partial sum overflowed the edge of the array.)

Since the processors are Boolean processors, we have to do the addition bit-by-bit. Let us describe the operations required at one of the processors:

Let **a** be the number originally at the point, **propag_in** the number propagated in, **propag_out** the number propagated out, and **b** the result to be stored (actually equal to **propag_out**, but given a separate name for clarity). Denote the ith bit of **a** by a(i) (with a(0) = lsb of **a**), and so on. Assume we are working to a precision of p bits. The basic algorithm is

$0 \rightarrow$ **carry**

FOR i = 0 TO p-1 DO

   a(i) @ **carry** @ **propag_in**(i) $\rightarrow$ **propag_out**(i), b(i)

   a(i) . **carry** + (a(i) + **carry**) . **propag_in**(i) $\rightarrow$ **carry**

With a little rearrangement, this becomes an algorithm to calculate partial sums along a line.

Algorithm sum_line

$0 \rightarrow$ **carry**

FOR i = 0 TO p-1 DO

    a(i) @ **carry** @ **propag_in**(i) $\rightarrow$ **propag_out**(i), b(i)

    a(i) . **carry** + (a(i) @ **carry**) . ¬b(i) $\rightarrow$ **carry**

This is executed simultaneously at each point in the array, with the propagation going in one direction only (e.g. left to right), initialised to 0 at the edge of the array. The algorithm requires only p global propagation instructions (and 3p pointwise operations), because the sum ripples across the array in one machine operation. The pointwise operations are needed to calculate the carry at each step. Note that p need not be any larger than the number of bits required to hold the largest number in the result (or the input), since we can just loop until **carry** is 0 everywhere and all the input planes have been used. Another potential refinement is to test for a(i)@**carry** being 0 everywhere. If it is, then b(i) must also be 0 everywhere, so that no global propagation instruction is needed in that iteration. Thus labelling the points becomes

Algorithm label_points

Let **marker** be the plane of marked points, then

1. Sum along rows from left to right – put the result in **horizlabel**.

$$\textbf{horizlabel}_{xy} = \sum_{k=0}^{x} \textbf{marker}_{ky}$$

2. Note the values at the right-hand edge of the array, propagate these values back across the array and put the result in **rowsum**.

$$\textbf{rowsum}_{xy} = \sum_{k=0}^{X-1} \textbf{marker}_{ky}$$

3. Sum **rowsum** along columns (from the bottom), and
    put the result in **vertlabel**.

$$\textbf{vertlabel}_{xy} = \sum_{1=0}^{y} \textbf{rowsum}_{x1} = \sum_{1=0}^{y} \sum_{k=0}^{X-1} \textbf{marker}_{k1}$$

4. Subtract **horizlabel** from **vertlabel** – giving the
    required result, **pointlabel**.

$$\textbf{pointlabel}_{xy} = \sum_{1=0}^{y} \sum_{k=0}^{X-1} \textbf{marker}_{k1} - \sum_{k=0}^{x} \textbf{marker}_{ky}$$

$$= \sum_{1=0}^{y-1} \sum_{k=0}^{X-1} \textbf{marker}_{k1} + \sum_{k=x+1}^{X-1} \textbf{marker}_{ky}$$

Note that the point labels will run from 0 to one less than
the number of marked points, without any gaps. If we wish
to reserve 0 as a label for the background, it is straightforward to adjust the routine above to generate labels
starting at 1, or we could just add 1 to the result.

Using the point labels to label the objects

We now have distinct labels on at least one point of
each object, and wish to use these to label the objects.
This can be done quickly by using the max within objects
routine which was introduced in section 3.2. However, in
the hope that it will give smaller labels, we will use a min
within objects algorithm, based on the same idea.

Algorithm min_within_objects

Let **image** be a binary plane containing the original
image, **marker** be a binary plane containing the selected
points, **ptlabel** be a bit-stack containing the labels for the
marked points, and **minlabel** be the bit-stack for the result.
Assume that **ptlabel** has p-bit precision.

1. Set i (the loop counter) to p-1.

2. Send a signal from all marked points at which
    the ith bit of **ptlabel** is 0, allowing the signal
    to propagate only at points within the objects
    (i.e. non-zero points of **image**).

3. Set the ith bit of **minlabel** to 1 where **image** is
    1 and a propagation signal has <u>not</u> been
    received. (This means that the result is 0 out-
    side objects, and 0 inside if there is a marked
    point within the object whose ith bit of **ptlabel**
    is 0.)

4. Delete from **marker** all points which received a
    signal, and whose ith bit of **ptlabel** is 1.
    (That is, delete points which are greater than
    the minimum so far.)

5. Subtract 1 from i, and if i is non-negative, go
    to (2).

Note that this algorithm not only sets the points inside
objects to the minimum label value, but also sets the points
outside the objects to 0, and leaves in **marker** only the
points at which the minimum was originally to be found.
This means that it leaves exactly one point per object in
marker if the original labels were distinct. Now we can
summarise the complete object labelling algorithm:

Algorithm label_obj_1

1. Select at least one point in each object in the
    image, for example, by using the points at which
    the mask

        0 0 *
        0 1 *
        0 * *

    fits.

2. Use the label_points routine to label the
    selected points.

3. Use the min_within_objects routine to pick one
    of the marked points in each object, and give
    its label to each point of the object.

The time taken to execute this algorithm depends primarily
on the number of points selected in step (1), since that
determines the number of bits needed to label the points.
It also depends to some degree on the distribution of the
points in the array, since the point labelling algorithm
takes longer if the points are not evenly distributed
between the rows.

Let P be the total number of points selected, and R be the maximum number of points in any one row. Then a straightforward implementation of the algorithm requires (2 log P + 2 log R) global operations, and about three times that many pointwise (or local) operations.

We would rather have a simple expression giving the number of operations in terms of the number of objects, N, or the size of the array, but both P and R also vary with the type of objects being labelled, so we will just look at a couple of special cases.

1. The best case occurs when only one or two points are picked in each object, and those points are evenly distributed over the array. This means that $P \approx N$, and $R \approx \sqrt{P}$ (in a square array), so that the number of global operations required is about 3 log N.

2. The worst case occurs when a very large number of points have been selected, either because there is a very large number of objects in the image, or because the objects have very awkward shapes, or both. As the points are selected using the mask

$$
\begin{array}{ccc}
0 & 0 & * \\
0 & 1 & * \\
0 & * & *
\end{array}
$$

    we cannot have picked more than XY/4 points in any X x Y image, and there cannot be more than X/2 points in any one row. Thus $P \leq XY/4$, and $R \leq X/2$, so the maximum number of global operations required is about (4 log X + 2 log Y).

We can put these figures into perspective by noting that a binary array processor needs at least log N operations simply to write out the label values, since these must have at least log N bits.

We can put a more stringent lower bound on the number of operations required by considering what data must be passed to each processor, in some fashion, to enable it to produce the label value. This data must carry two sets of information: information which will ensure that the objects have distinct labels, and information which ensures that each point in a label has the same value.

In order to ensure that the label on each object is distinct, at least one point in each object must receive log N bits of information encoded in some way. This data must be unique to each object, so the most economical way of passing it to each object would be log N global propagation instructions. (Note that using positional information, such as the x,y co-ordinates of points in the object, does not get round this, since we are effectively passing (log X + log Y) bits of information to each point when we generate the x and y co-ordinates.)

To ensure that the labels on each point of the object are the same, we need to pass log(label value) bits from one point of each object to all the other points of the object. Again, as the objects can be of arbitrary size, and we need to pass different data to different points, log N global propagation operations is the most efficient way of passing the information to each point. Note that we can detect that labels on different points of an object differ using only log N local propagation instructions, but we need log N global propagation instructions to fix the labels if they differ.

We cannot combine these two sets of operations since, in the first case, information must be sent between objects and, in the second case, the information must only be passed within the objects, so that the propagation must be constrained differently in the two cases. So, in general, we need at least 2 log N global propagation instructions (or some equivalent) to label N objects on CLIP4. label_obj_1 requires about 3 log P global operations; and for most images P is comparable with N (since only a few points are picked in each object), so label_obj_1 is not far from this lower bound.

It is impossible to prove that, for most images, P is comparable to N. However, it is possible to adduce two arguments in support of the idea. Firstly, if the objects have a fixed maximum size, then there is some number k such that $P \leq kN$. Thus $P \sim N$, and $\log P \leq \log N + \log k$. Secondly, for the case where the objects do not have a fixed maximum size, let us consider an example.

Figure 2.5 shows an early stage in segmenting an image of a diatom, which we wish to improve by merging small regions and deleting isolated lines [17]. An essential step in doing this is to label the regions. There are 46 (4-connected white) regions in this image, so at least 12 global propagation instructions are required. Simply using the mask

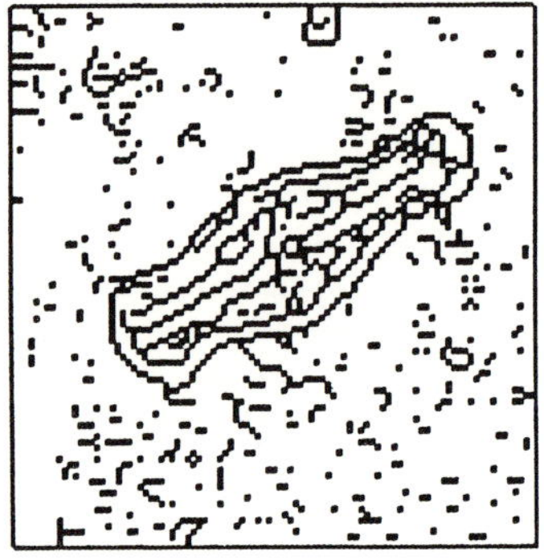

Figure 2.5   Partially segmented diatom image.

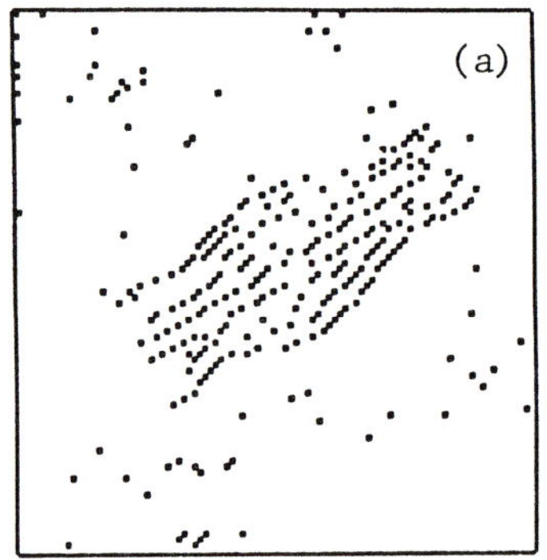

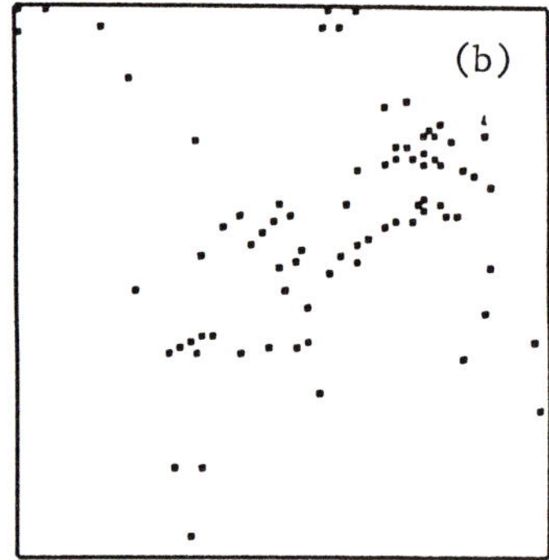

*Figure 2.6   Points selected from Figure 2.5:   (a) only
using a mask;   (b) using a mask and a global propagation
instruction.*

```
* 0 *
0 1 *
* * *
```

(the 4-connected analogue of the mask suggested above),
selects 274 points (see Figure 2.6(a)), so 26 global propagation instructions are required. In the 8-connected case,
the appropriate mask excludes more points, so the performance of the algorithm is slightly better.

The main potential for speeding up this algorithm lies
in improving the way in which the points are selected. The
ideal would be to select only one point in each object but,
as we saw in section 3.2, this is not easy. It is not made
any easier by requiring that the extra time taken to select
the points should be outweighed by the consequent speed-up
in the rest of the labelling. However, we can sometimes
reduce the number of points significantly, using quite simple operations. For example, we can adapt the
top_left_point algorithm in section 3.1. This gives us a
means to select at least one point in each object.

Algorithm select_points

1. Select the points using the mask

```
0 0 *
0 1 *
0 * *
```

    as before.

2. Send propagation signals from each selected
    point and receive them in directions 2-5, pro-
    pagating only within the original objects, and
    delete all points which receive a propagation
    signal.

Step (2) will not delete all the points in an object, since
the point which started the signal will remain. If the
object is convex, only one point will remain, but unfortunately this is not usually true for objects with concavities, since the concavities may prevent the signal from
higher points in an object from reaching the lower points.
(For similar reasons, the global propagation used in the
first step of top_left_point can be reduced to the local
mask application which we used as the first step here.)

on very irregular objects.   To give an impression of the
improvement, Figure 2.6(b) shows the points selected from
Figure 2.5 using this technique and these should be compared
with the result in Figure 2.6(a).   The number of points
selected has been reduced from 274 to 79, so that the number
of global operations needed for the rest of the labelling is
reduced to 20.

4.2 A Fast Consecutive Labelling Algorithm

One drawback of label_obj_1 is that the labels do not
run from 1 to N, but lie somewhere in the range 1 to P.
Since P is typically of the same order as N, this does not
usually matter; but if we wish to guarantee consecutive
labels, all we have to do is ensure that we select unique
points on each object, and then use the same point-labelling
technique that we used in label_obj_1. To get the unique
points, we can either use the x,y co-ordinate method of section 3.2, or use the points which are produced as a by-
product of the previous labelling method.   The latter is
often slightly faster, because the intermediate point labels
which it uses are smaller. This gives us

Algorithm label_obj_2

1. Select a unique point in each object by

    (a) selecting the points at which the mask

                    0 0 *
                    0 1 *
                    0 * *

        fits, which gives us at least one point per
        object;

    (b) using the label_points routine to label the
        selected points; and

    (c) using the min_within_object routine to pick
        exactly one of the marked points in each
        object.

2. Use the label_points routine to label the
    selected points.

3. Use global propagation to spread the label from
    the labelled point in each object to all the
    other points. For a small overhead, we could
    simply use the min_within_object routine to do
    this spreading.

Clearly this algorithm will take roughly twice as long to
execute as label_obj_1, since it is effectively label_obj_1
executed twice, in a suitable fashion.

5. SOME EXECUTION TIMES

We have already discussed how the various algorithms
compare in reasonable detail. However, we will give a rough
summary by listing the times (in ms) required to count or
label the 4-connected white regions in Figure 2.5, the partially segmented diatom. Note that the times required for
other inputs may not be proportional to these times, because
the times for the different algorithms depend on different
features of the image.

Algorithm	CLIP4A	CLIP4D
Counting objects		
Sequential removal	90	4.2
Reduction to points	19.5	1.2
Connectivity number	3.2	0.3
Labelling objects		
label_obj_1	18.5	1.3
Improved version	15	1.1
Consecutive labelling	26	2.1

Note that by adapting the object labelling of section 4.1,
the reduction to points routine can be speeded up to about
15 ms and 1.1 ms respectively.

6. DISCUSSION

Communication between processors is often a limiting factor in computers with many processors. Two-dimensional arrays used for processing images are relatively well-served: much of the computation only requires data from a small neighbourhood, so merely connecting each processor to its immediate neighbours is adequate; this simplifies the hardware greatly. However, a problem still exists when we wish to send data long distances across an array. Many tasks (probably most, even in image processing) need to do this at least occasionally. In the conventional cellular array, without special provisions, it takes at least d operations to transfer data a distance d through the array. This is often a major overhead.

Global propagation is often an efficient way of speeding such data transfers. It also allows much more efficient use of the processors in some cases, such as in the sum_line algorithm. When combined with an 8-connected array, it is especially suitable for image processing algorithms, because it enables one to exploit the similarity in structure, such as connectivity, between images and the array. For some examples of uses of global propagation unrelated to the connectivity of images see [23].

In this Chapter we have seen several examples of algorithms which are fast because they can quickly send small quantities of data large distances across the array. Without global propagation, or something similar, both counting and labelling objects would require O(size of array) operations.

CHAPTER THREE

SOFTWARE FOR CLIP4

If image processing systems are to be successfully applied to real-world problems, three things are required - a hardware system. a software system and a set of algorithms or operational primitives. It is the software system which provides the interface between machine and user and is therefore the public face of the whole system. This Chapter comprises descriptions of the two principal means by which the CLIP4 system can be programmed.

The first of these, already mentioned in Chapter 1, is the assembly language CAP4, developed by Alan Wood and described in his thesis 'Parallel Processing Techniques for Image Sequence Analysis'. Although this level of language is ideal for, amongst other things, the exploration of algorithm construction, it was realised that a higher-level language would be required for application programming. After experimenting with a number of alternatives including PASCAL, David Reynolds and Paul Otto developed IPC, an extension of the C language running under the UNIX operating system. The bulk of the description here is taken from David Reynolds' thesis 'Automatic Generation of Image Segmentation Procedures for a Cellular Array'.

CELLULAR LOGIC
IMAGE PROCESSING
ISBN 0 12 223330 1

1. INTRODUCTION

The prototype CLIP4 can be used as a stand-alone computer once its programs have been developed and punched on paper tape. However it lacks the I/O facilities which would be needed for developing programs, such as a keyboard interface, or disc storage; thus another (conventional) computer is also needed for program development. The computer currently used to support CLIP4 programming is a PDP-11, running the UNIX operating system. This combination, together with the PDP-11's peripherals, provides a very good environment for developing CLIP4 programs and other software.

CLIP4 is connected to the PDP-11 as though it were a peripheral device, such as a disc; this means that it is easy, from the PDP-11, to load and run CLIP4 programs, or to transfer data between the two machines. Thus, all the facilities of a general-purpose computer system are available to CLIP4 users, and comparatively little software has to be developed specifically for CLIP4.

There is one assembly language for CLIP4, CAP4. This is a straightforward symbolic assembler, without any macro facilities; one line of CAP4 code corresponds to one CLIP4 instruction. To go with this assembler, there is a simple linker, and a utility called CRISP which provides the run-time support for CAP4 programs; this includes simple debugging and I/O facilities. In addition, there is a library of standard routines, mainly for handling grey-level images, called CAPLIB.

As with any other assembly language, CAP4 reflects every detail of CLIP4's architecture, and allows very efficient programming. In particular, it is often much more efficient for binary image processing than the current implementations of the higher-level languages for CLIP4, mainly because these implementations drive CLIP4 from the PDP-11, and tend to be slowed down by the current PDP-CLIP4 interface. On the other hand, CAP4 becomes very cumbersome to use for large or complicated programs. This is aggravated by the lack of macro facilities, because the limited instruction set of the CLIP4 controller means that several instructions

are needed even for simple operations like accessing memory
indirectly, or rotating the direction lists used in local
array operations.

When writing large programs in CAP4, another, more
unusual, difficulty is encountered: allocating the image
memory. Because there are only 32 planes of memory in
CLIP4, it is not feasible just to assign a set of planes to
an image variable for the entire duration of a program;
memory must be reused whenever possible. This allocation
becomes particularly onerous when grey-level images are
being handled, since they are usually stored in the minimum
number of bit-planes in order to conserve memory. This
means that an image may occupy 6 planes of memory in one
section of a program and 10 in another. In such circumstances, efficient static memory allocation becomes
difficult or impossible.

2. THE CAP4 ASSEMBLY LANGUAGE

The primary job of an assembler is to reflect the
hardware structure as defined by the machine-executable
instructions, permitting the use of the hardware facilities
to the full. This could be done by writing instructions in
a numerical format (octal, decimal, binary, etc.). However, this is impractical from a user's point of view and so
the secondary task of an assembler is to enable a user-
friendly language to be used, translating from this into
machine code. These two requirements are somewhat contra-
dictory since the more user-friendly the language (which
tends to mean higher-level), the less it reflects the
hardware structure.

This dichotomy is generally resolved by defining an
assembly language such that each statement should represent
one machine instruction, but allowing values (addresses,
constants, instruction codes, etc.) to be represented by
symbols, some of which are user-definable. The entities so
represented will be limited to those which the hardware
regards as single operable units, say 16-bit integers or
arrays of 96 x 96 bits. Operations on other entities
require sequences of instructions defined by the programmer.
An assembler may thus be best described in terms of its
handling of instructions and symbolic data.

The rest of this section is divided into four parts, the first three dealing with those instructions which are peculiar to CLIP4 as opposed to the rest of the instruction set which is mostly conventional in nature, and the fourth describing the assembler directives. For full details of the current instruction set and the general format of the program text, see the CAP4 Programmer's Manual [24].

In this section, **emboldened** symbols are literals in the language (terminal symbols). All other symbols should be taken as denoting appropriate terminals. The language makes no distinction between lower and upper case letters.

2.1 Array Instructions

The heart of the CLIP4 machine is the processor array which is analogous to the ALU of a conventional computer. It may be set to perform any one of a large number of possible functions of its operands by means of the **set** instruction. The operands to be used are loaded from the bit-plane storage (D-levels) into the A- or B-registers as appropriate, by means of the **lda** and **ldb** instructions respectively. The operation is initiated and the result stored into a D-level with the **pst** (process and store) instruction.

Set

This instruction is the counterpart of the array in the assembly language: it is the means by which the user specifies the array function. Since the number of possible operations is very large, the format of **set** is necessarily complex. However, an attempt has been made to keep it as logical as possible with the proviso that, in the case of program language formats, logic tends to be subjective. Another reason for the complexity of this instruction is that it is effectively a microprogramming instruction. By this it is meant that, whereas an instruction such as add in a conventional computer has an op-code which is decoded by the hardware to produce the control signals required by the ALU to perform an addition, with the **set** instruction these signals are more nearly represented by the instruction's operands. It is thus a lower-level instruction than others in the language.

The **set** instruction format consists of three fields, separated by commas, the first of which is compulsory:

set output-function, neighbourhood-part, default-options

The output function specifies the Boolean function of the A-register and P input which is to produce the output result to be stored in a D-level on completion of the operation.   This is specified in terms of the symbols a and p along with the operations + (OR), . (AND), @ (EXclusive OR), − (NOT) and parentheses.   There is no restriction on the combination of symbols except that a and p may only appear once each in the expression.   This is no real hardship since all Boolean functions can be represented in this way. In addition to the above, the symbols 0 (zero) and 1 (one) may be used to signify an all-clear and an all-set output respectively.

The p symbol refers to the P input of the processor cell (Figure 1.1) which is a combination of the B-register input, the C-bit and the neighbourhood gating.   Thus the p symbol refers to the neighbourhood part of the instruction.   However, when the P input is identical to the B-register input, (i.e. when a simple Boolean function of A and B is being performed), it would be untidy to have to specify in the neighbourhood part that B is to be used (see later) and so the assembler automatically sets the bit in the instruction which enables the B input.   Thus an AND of two bit-planes would be written

set   a.p

The neighbourhood part of the set instruction is split into two sections

[P-set] propagation-function

The propagation function is identical in format to the output function.   It specifies the function of A and P which will provide the N output from the processor cell.

The P-set, which is delineated from the propagation function by square brackets, specifies those signals which will form the P input to the processor.   The first of these signals is the set of neighbouring cells whose N outputs will be selected by the input gating.   This list is formed by specifying a string of digits (from 1 to 8) signifying the relative direction of the required neighbour according to the convention for directions.   If a sequence of directions is needed, then the form digit-digit is allowed, e.g. for all directions in square tessellation, 1-8. The directions list may be contained in a general-purpose register [24] which is indicated by inserting %reg (where reg is the number of the register to be used) in place of the set of directions in the P-set.

The other elements in the P-set are letters indicating that the B-register input should be enabled (**b**), the C-bit input should be enabled (**c**) and whether the edges of the array should be set to 1 (**e**).

The final field of the **set** instruction is a string consisting of a subset of the letters {**e c r h**}. The uses of e and c are equivalent to those in the P-set above; they may be placed in the default-options field if preferred. The **r** switch enables the ripple-carry gate in the processor cell to produce a carry in bit-column or bit-stack arithmetic. The **h** switch sets the array into hexagonal connectivity (the default is square). In hexagonal mode, the direction list should only contain directions 1 to 6 inclusive.

It can be seen from the above description that the **set** instruction specifies a wide variety of functions. These may be divided according to two orthogonal classification schemes viz. number of operands and extent of propagation. Table 3.1 gives examples of set instructions classified according to these two methods. The operand classification is self-explanatory. The classification by propagation is determined by the neighbourhood connectivity and the type of propagation function:

1. A null propagation instruction is one in which the states of a cell's neighbours are not included in the propagation input. Thus both zero operand instructions and all Boolean operations of one or two operands are null propagation.

2. The local propagation instructions are determined by two properties. Firstly, the propagation input to the cell (p) must be a function of the neighbours' states (i.e. there must be a direction list specified) and secondly, the propagation function must not be a function of the propagation input (i.e. p must not appear in it). Thus information may be received from the neighbourhood, but the signal sent to the cell's neighbours does not depend on this information.

Table 3.1   **set** instruction examples

Propagation	Number of Operands		
	zero	one	two
null	set 0 (clear a bit-plane)	set -a (invert)	set a.p (AND 2 bit-planes)
local	set -p,[678]1 (set bottom left point)	set p,[6]a (shift up)	set -p,[b%mask]-a (used in mask fitting)
global	set p,[6]-p,e (alternating horizontal stripes)	set a+p,[26]a+p (spread image vertically)	set a.p,[1-8b]a.p (labelling)

3. Global propagation occurs when the conditions
   for local propagation exist except that the propagation output (the N processor output) is a
   function of the propagation input. This gives
   rise to the possibility of information from one
   processor being transmitted to all other processors in the array.

The main practical difference between these **set** classes is
that the null and local propagation instructions take the
same time to execute and have a constant speed (the basic
operation cycle). Global propagation instructions, on the
other hand, have a variable speed which is data-dependent.
This is because a global propagation signal, once started,
might travel a distance of one cell or many cells. There
is a hardware mechanism which is intended to detect the completion of a global propagation (and signal the end of the
instruction at that point), by noting when the propagation
outputs from all processors have remained stable for two
cell propagation cycles.

lda, ldb and pst

In order that bit-planes may be operated upon by the
array (as specified by the **set** instruction) they must first
be loaded into either the A- or B-registers (**lda**, **ldb**
respectively), and then the process must be initiated, storing the result on completion (**pst**: process and **st**ore).
These three instructions have the same format.

The operand field in these instructions specifies which
D-level is to be accessed. In the case of the two load
instructions the D-level will be read from; the **pst** D-level
will be written into. Two forms of D-level specification
are available: absolute and indexed. In the absolute
case, a number or previously defined symbol is given which
represents the address of the D-level required. This must
be within the range of available addresses, which is 0 to 31
for CLIP4, and the assembler will give an error indication
if this is exceeded.

Indexed addressing is in the format:

$$d(r)$$

where d is a D-level and r is a register indication. The
action here is for the contents of register r to be added to
d to form the address of the D-level for the instruction.
If there were no d specified, then it would be taken as
zero. The result of this addition must be in the range

mentioned above, although there is at present no hardware check on this and an error cannot be detected at assembly time. The main uses of indexed addressing are for iteration of instructions over a stack of D-levels as in the case of bit-stack arithmetic and for passing D-addresses as parameters to subroutines.

bzea, bnza

After each **pst**, i.e. when an array operation has been performed, a condition code is set according to whether the result was an all zero bit-plane or not. This code is used by two conditional branch instructions which take the branch if the result was zero (**bzea**) or non-zero (**bnza**). Although only one condition is tested on the array, other conditions may be emulated by means of a sequence of instructions prior to the branch.

count

It is sometimes necessary to extract information from the array for processing in a conventional serial form. An example would be the measurement of the area of an object. The **count** instruction takes the bit-plane currently held in the A-register and counts the number of bits set to 1. The result is stored in general register 14. Thus, any property of a binary array that may be expressed in terms of a quantity of bits can be extracted by means of **count**, e.g. the circumference of an object, the x,y co-ordinates of a point, the distance between points, height of a histogram, etc.

This instruction causes a destructive read-out of the A-register, since it is linked to the mechanism for input/output (see next section). Therefore if the data should still be required, it must be reloaded.

2.2 Input/Output Instructions

Two methods are available for moving pictorial data into and out of CLIP4. There is an interface to the PDP-11 which enables any form of binary data to be transferred to and from CLIP4 (e.g. from disc or tape files), and there is a dedicated I/O unit for data to be displayed and gathered from a standard television signal (e.g. from a TV camera or video recorder). The use of the dedicated I/O system will be described here.

Figure 1.3 shows the I/O system in relation to the rest of the system and will be referred to in the following. Also, when a TV camera is mentioned, it is taken to mean any device producing a TV-compatible signal.

scan, scans

The first step in getting any pictorial data into CLIP4 is to cause the image to be loaded into the input memory. The **scan** instruction effects this transfer by taking a single frame of the TV image, digitising it to 64 grey levels (6 bits) and writing each bit-plane into the 6-plane input memory stack. This will not then alter until the next **scan** (or **scans**) instruction.

The **scans** instruction copies the output memory into the input memory. This enables a previously displayed picture to be re-input or allows the output memory to be used as an additional 6 bit-planes of slow memory if needed. Neither **scan** nor **scans** takes any operands.

ipa, ipas and derivatives

The basic bit-plane input instruction is **ipa** which has two operands: the mode of input, and an associated value which is interpreted according to the mode. However, although this instruction is available to programmers and reflects the structure of the underlying machine-code, it was decided to split **ipa** into four logical instructions each corresponding to one of the four modes of **ipa**. This has meant that the somewhat anonymous modes have been replaced by more memorable mnemonics.

The **ipas** instruction causes the A-register to be loaded from the contents of a buffer situated in the CLIP4 program memory. This is the means by which data is acquired from a computer interface.

The four input instructions are as follows, each taking one operand. However, the hardware (and consequently the assembler) has no facility for indexing this operand.

> **lth and hth**   These two instructions load the A-register with a bit-plane whose pixels are set to 1 according to whether their grey value is higher than, or lower than the threshold value given in the operand field (**hth** and **lth** respectively). Obviously, the previous contents of the A-register are lost, but there is no effect on the input memory, so a subsequent **hth** or **lth** can be used to give a

differently thresholded input. The operand value can be overridden at run-time by a thumb-wheel switch on the CLIP4 console.

pln This gives access to the input memory plane by plane. The operand of the **pln** instruction specifies the plane of the input memory (0-5) which is to be copied into the A-register. Using repetitions of this instruction, full grey-level pictures up to 6 bits in precision may be passed to a set of D-levels. However, lack of indexing on the operand makes this process less convenient than it could be.

slc Some applications require that all pixels in a grey-level image with a particular value be detected. The **slc** (slice) instruction allows this. An example would be to create a histogram of the grey-level distribution of a scene, in which case a **slc** operation would load the A-register with each slice in turn.

opa, opas

The last of the I/O instructions is the pair for output and display. Both **opa** and **opas** dump the A-register (destructively) to a bit-plane in the output memory, as specified in the operand field. The output is continuously displayed by constructing the grey level of a pixel from the bit-stack formed by the output memory.

In addition to performing the same operation, **opas** dumps the A-register to a buffer in the program memory. This would be used either to allow an attached computer access to pictorial data or to act as a further single bit-plane of memory which may be used with **ipas**.

2.3 External Communications

A general facility is provided for gaining access to an attached computer or other device without needing knowledge of the nature of the device. This is the **hi** instruction (halt-interrupt) which simply halts the CLIP4 processor and sends a signal to the interface with the attached device. In the current system, the interface generates an interrupt on the PDP-11 bus. After the signal has been generated, the external system has control and must make its own arrangements for the transfer of information and restarting CLIP4.

The **hi** takes one operand in the range 0 to 4095 which is ignored by the CLIP4 controller. However this operand appears in the lower 12 bits of the machine instruction and so may be examined by the external device or interface.

2.4 Directives

In addition to instructions which produce executable machine code, an assembly language will have a number of other statements which direct the assembler to take some action. The directives in CAP4 are of two types: for storage allocation and information, and are distinguished from instructions by having a full-stop as the first character of their names.

3. IPC: A HIGH-LEVEL LANGUAGE FOR CLIP4

High-level languages have many advantages compared with low-level languages: a program written in a high-level language is much easier to write, modify and maintain than one written in assembly language and is more portable; on the other hand, it is usually possible to write a more efficient program in assembler than in a high-level language.

For algorithm and program development, the advantages of a high-level language far outweigh the potential disadvantages unless the language is extremely inefficient; thus it was very desirable to develop a high-level language for CLIP4.

Unfortunately, the restricted image memory of CLIP4 means that a small inefficiency in the use of memory can cause a very large speed reduction. For example, if a program uses 33 planes of image memory, rather than 32, then it needs to use disc memory which is about 5000 times slower than the CLIP4 image memory, so that the program may be several thousand times slower. This order of slow-down is unacceptable for any significant task, even when developing programs. Thus one of the primary goals for implementing a high-level language on CLIP4 was that it should make, or should allow, very efficient use of image memory.

The construction of a high-level language for CLIP4 requires more than just the implementation of a standard language on a new computer. The data formats and operations provided by a cellular array are completely different from those of a conventional serial computer and it is important to reflect this in the language. For simple architectures, such as vector processors, it is possible to automatically detect parallelism by examining indexed looping structures (such as the FORTRAN DO loop) and so avoid having to define special language extensions for parallel execution. Whilst useful for program portability such automatic transformations are not easily able to make use of the significantly different properties of cellular arrays. Indeed experience has shown that, even when available, such methods are unsatisfactory for users and that explicit parallelism in a language is preferable. It is thus necessary to consider what special data structures, control structures and primitive operations are fundamental to cellular image processing languages. These have been embodied in an extension of the language C, referred to as IPC (Image Processing C).

3.1 The Image Data Type

The single most important characteristic of a bit-serial cellular array is that it manipulates, as fundamental objects, binary arrays. In the case of CLIP4 these are stored in the machine in the CLIP4 D-levels. IPC provides a data type called **image** which is used to represent integer arrays. Variables of type image are treated by the language as single objects, not as collections of pixels. Physically an image variable consists of a small descriptor block in the host computer which normally has attached to it corresponding image space in the cellular array. Images are declared in the usual way, for example

image im1, im2;

declares im1 and im2 as image variables able to represent bit-stacks in CLIP4. The physical CLIP4 space is only attached to the variables at run-time. The image assignment statement will allocate space to an appropriately initialised image variable as necessary and this allocation will be varied by other assignments when required. In addition functions are provided for explicitly de-allocating images (reclaiming their CLIP4 space) and for dynamically

modifying image lengths.   For example, if im1 and im2 are
8-bit images then

$$aadd(im1, im2)$$

which adds the contents of im2 into im1, will extend im1 to
9 bits temporarily.   Then, if the data is such that the
ninth bit-plane carries no information, im1 will be
compressed to the minimum size needed to hold the result.
To explicitly modify an image's length a variety of routines
is provided, for example

$$around(im1, s, n)$$

reduces (or increases) im1 to n bits by discarding its least
significant s planes and discarding (or adding) most significant planes to achieve the correct length.   The partial
ordering of pixel values is preserved by clipping pixel
values which lie outside the range expressible by an n-bit
number.   Images can be explicitly deleted by

$$dump(im1, im2);$$

This illustrates one minor feature of IPC, that of variable
length parameter lists for functions, like dump, which simply process each argument image in turn.

When images are allocated space (by assigning them the
result of some image function or by an explicit 'create'
call) then two possible problems can arise.   Firstly, there
may not be sufficient free D-levels in CLIP4 for the image
requested, in which case one or more of the current images
are swapped out on to a disc-based temporary file, thus
freeing more CLIP4 space. Once sufficient total free space
is available in CLIP4 then the image allocation can proceed.
However, IPC maintains compatibility with CAP4 subroutines
by assigning images contiguous bit-stacks in CLIP4.   Thus
as the space becomes fragmented it may be necessary to run a
garbage collector to repack the images in CLIP4 leaving a
contiguous block of free space.

These two store management functions normally are
completely transparent to the user but functions are available to provide explicit control of disc swapping, if needed
for efficiency.   Firstly, images can be explicitly swapped
out by

$$deactivate (im1);$$

when it is known that space will be needed and a particular image will not be used for a while. This makes no difference to later use of the image; images are swapped in when needed. Secondly, images can be explicitly locked into CLIP4 to prevent them being swapped out for the duration of some critical piece of code. Finally, images can be tagged as READONLY which indicates that they will not be modified so that their disc copy will remain valid and they need not be physically swapped out. This is used for images such as input data which are not to be changed.

In addition to the dynamic length modification, images can be associated by **subimage** links, so that one image corresponds to a subset of the bit-planes of a second image. The association is part of the allocation system because, as images are relocated by the garbage-collector or swapped by the disc pager, the subimage links have to be preserved. Subimage structures are useful in special circumstances to save D-level space but are rarely needed.

Finally, the image variables have a number of attributes associated with them, most of which are maintained by IPC but are user-accessible. These attributes are:

1. Details of the current CLIP4 space allocated to the image, in particular its length.

2. Details of any current disc swap space allocation.

3. A set of flags giving the image subtype – permanent, temporary or protected temporary.

4. A set of flags for controlling the disc swapping – LOCK and READONLY.

5. A **sign** flag to indicate whether the image represents signed or unsigned integers. All main IPC subroutines can handle arbitrary combinations of signed and unsigned images with appropriate sign extensions where needed. In addition, when an arithmetic subroutine yields a signed result whose sign plane is blank then the space taken up by that plane is reclaimed and the result set to unsigned. The maintenance of the sign flag is completely transparent to the user.

As well as being used in the internal image descriptor, most of these attributes are also attached to any disc copy of the image by the image file access primitives.

So images can be written to a file and read from a file as single objects and lengths and sign information will be preserved.

3.2 The IPC Functions

All processing of images is carried out in IPC by use of nested function calls and an image assignment statement. Image assignment is denoted by $= as in

```
im1 $= add(im2, im3);
```

In this example the pointwise sum of im2 and im3 will be placed in a temporary image of suitable length. If im1 has not previously been allocated space then the temporary image space is just attached to im1; if im1 has been allocated space then it is reclaimed first. In a case such as

```
im1 $= add(im1, im2);
```

this arrangement uses unnecessary space for the temporary sum; to cater for this, many IPC functions come in two versions. One version, e.g. add, is a non-accumulator function which returns a separate, newly allocated, result in a temporary workspace to be assigned by $= to the target image. The other format is that of an accumulator function, e.g.

```
aadd(im1, im2);
```

which overwrites its first argument with the generated result and minimises the workspace required.

Function calls can be nested as in

```
im1 $= sub(im2, medn4(im3));
```

In this example the image im3 is median-filtered by the medn4 function and the result is returned in a separate temporary image (im3 is unchanged). This temporary image is then subtracted from im2 to yield a second temporary image which is then assigned to im1. The first temporary has to be reclaimed at some stage; this is done by the function that uses or `consumes´ it. All IPC library functions, before returning their result, will delete (and reclaim the space from) all their arguments which were temporary images. Thus the results returned by non-accumulator functions are consumed by the first function to use them.

This arrangement allows efficient use of CLIP4 D-level
space; without it, virtually all nested function calls
would cause disc paging to be used.

The accumulator routines also return a pointer to their
result to allow them to be nested in the same way as the
non-accumulator functions.   For example

```
iml $= sub(im2, amedn4(im3));
```

gives an identical final result to that of the previous
example except that im3 is replaced by its median-filtered
version and no extra temporary space is required to hold the
filtered image.

The set of functions actually provided is fully
described in [17]. In addition a system of shared personal
libraries has been established to encourage users to make
any specially developed functions of general interest available to all users.   Currently about eighty fully documented
user subroutines are accessible in this way and include
functions for image rotation and stretching, segmentation
primitives (various thresholders, edge detectors and postprocessing functions) and morphological operations (such as
the watershed algorithm).

It is difficult to convey the ease or difficulty with
which any particular algorithm can be encoded in IPC but, in
an attempt to do so some examples of short programs are
given here.

Example 1 – Find the centroid of an object

Suppose we have a binary picture (im) of a single object
and wish to locate its centroid.   This is defined by

$$\bar{x} = \sum_i x_i / \sum_i 1 \qquad \bar{y} = \sum_i y_i / \sum_i 1$$

where i ranges over all pixels of the object.   Now $(\sum_i 1)$ is
just the area of the object found by

```
area = volume(im);
```

To generate $\sum_i x_i$ we first label each array point by its x
co-ordinate by

```
xcoord $= ramp(dirc[8], 7);
```

Next, those co-ordinate values corresponding to object pixels are summed by

```
xsum = volume(ands(xcoord, im));
```

When all packaging has been carried out, the complete program becomes

```
#include    <clip.h>            /* defines images, etc. */

main()
{
    image   im;                 /* declare variables */
    float   area, x, y;         /* floating point */

    openclip();                 /* initialises CLIP4 */
    setname(im);                /* initialise image */

    im $= inpics(1);            /* take picture from camera */
    area = (float) volume(im);

    /* Find the centroid by sum(x)/sum(1) */

    x = (float) volume(ands(ramp(dirc[8],7),im))/area;
    y = (float) volume(ands(ramp(dirc[6],7),im))/area;

    /* Formatted print of centroid co-ordinates */

    printf("Centroid of im is at %4.1f, %4.1f\n",x,y);
}
```

Example 2 – Delete small objects

Given a binary picture (im) containing one or more connected objects, delete all objects which are smaller than a given size.

If the size threshold can be specified as a minimal diameter then we can use

```
im $= labelim(im, shrinkr(im, diam));
```

where shrinkr is an iterated binary shrink function (8-connected) and labelim is a component labelling operation. If the size threshold is given as an exact number of pixels then the sizes of all objects must be measured exactly. This can be done by repeatedly selecting a separate object and measuring its size, as in

```
  out $= clear(1);
  forever {
      obj $= labelim(im, firstpoint(im));
                                  /* get next object */
      sizeobj = volume(obj);
      if (sizeobj > THRESHOLD)
          aors(out, obj);        /* keep large objects */
      else if (sizeobj == 0)
          break;                 /* exit if none left */
      aandnots(im, obj);         /* delete object */
  }
```

Alternatively, the efficient labelling algorithm developed by Otto (see Chapter 2) can be used.   Consider the following code

```
            int array[256];

            labeled $= region8(im);
            hista(labeled, array);
```

Here region8 assigns a unique integer to each 8-connected region of the image and hista places a histogram of these labels in the (serial) variable array.   The entry array[i] is then the number of pixels with label i (i.e. the size of the ith object assuming consecutive labels).   The desired regions can then be determined by scanning this serial array.

Example 3 – Constrained local averaging

One step of a simple region-merging algorithm is to smooth the grey levels of an image by local averaging but constraining the local regions not to cross a known set of object boundaries.   Let the grey image be im and the object boundaries be represented by thin 8-connected lines in the binary image edge.   Then the local sum can be found by

```
  image av, temp;

  av $= clear(1);
  temp $= aandnots(im, edge);   /* set pixels on boundaries
                                     to zero grey level */
  for (d = 2; d <= 8; d += 2)
      aadd(av, shift(temp, dirc[d]));
                                /* 4-connected local sum */
```

To produce the final local average each pixel in the sum
must be divided by the number of local neighbours after
being constrained by edge.   This is done by

```
image scale;

scale $= cntnbr(nots(edge), dir4));
                                /* scale(i,j) = number of
                                   pixels in constrained
                                   neighbourhood of (i,j) */
adiv(av, scale, 0);             /* pointwise divide of sum
                                   by local scale factor */
aandnots(av, edge);             /* suppress boundary regions
                                   from final average */
```

where cntnbr(im, dir4) is a 4-connected local sum of a
binary image.

Example 4 – Delete active lines

Given an 8-connected (thin) edge map representing an
attempted image segmentation, a useful postprocessing step
to clean up the segmentation is to delete all edge pixels
which are adjacent to only one region of the segmentation.
Such pixels are called active edges by Perkins [25].   The
first step is to label the segmentation regions by

```
image labeled, edge;

labeled $= region4(nots(edge));
```

where edge is the 8-connected edge map to be cleaned up.
Pixels in a particular region all receive the same numerical
label, the labels being unique to each region.   To detect
pixels adjacent to only one region we find the minimum and
maximum label in each 3 x 3 neighbourhood surrounding the
edge pixels and compare the two.

```
image maxl, minl;

aandnots(labeled, edge);        /* set borders low */
maxl $= maxn(labeled, alldirec); /* local max */

aors(labeled, edge);            /* set borders high */
minl $= minn(labeled, alldirec); /* local min */

aands(edge, cmp(minl, maxl));
```

The labels of the edge pixels themselves are set to 0 whilst
the local maxima are being calculated and to the largest
label value in the entire image during the local minimum
operation so that the values found are always those for the
regions themselves, not those for edges. If an edge pixel
is adjacent to two or more regions then since the regions
all have different labels, the local maximum and local
minimum of the labels (over non-edge regions) will differ.
However, if an edge pixel is adjacent to only one region
then the local extrema will be the same, cmp will set the
corresponding edge pixel high and aands will delete it from
the edge map.

3.3 Binary Masks and Direction Lists

As well as images, a fundamental data structure for a
cellular array is that of a local neighbourhood. For local
grey-level processing, such as convolution, the neighbourhood can be specified using the normal serial array structures of the base language. For example a subroutine for
3 x 3 convolution is

```
linkim
conv (im, mask, scalen)
image im;
int mask[], scalen;
{
    register int i;
    register linkim acc;

    prot(im);
    acc = clear(1); acc->ifg |= PROT;
                                /* get a temp */

    for(i=0; i<9; i++)
        aconvp(acc, shift(im,dirc[i]), mask[i]);

    unprot(im); release(im);
    ascale(acc, 1, scalen);
    acc->ifg &= ~PROT;
    return(acc);
}
```

where mask is a 9-element integer array specifying the kernel weights and

$$\texttt{aconvp(acc, im, int)}$$

calculates

$$acc_{ij} = acc_{ij} + int * im_{ij}$$

Direction Lists

The shift subroutine used above required a specification of which array direction (0 to 8) to shift along, so that a direction list data type is needed to represent a set of array directions. In IPC a direction list is just an unsigned integer holding a bit-map of array directions. To specify a particular direction two aids are available: the array **dirc[i]** gives the bit-maps for the single directions 0 to 8 and the function strdir converts a direction list specified in a string to a bit-map. For example

$$\texttt{dir = strdir("02-4");}$$

sets dir to represent directions 0,2,3 and 4. Thus, for instance, a 4-connected grey expand could be specified by

$$\texttt{exp \$= maxn(im, strdir("02468"));}$$

Masks

As well as the direction lists, a particular case of local neighbourhood which has proved useful is that of the local binary mask. In IPC a **masktype** data type is provided to represent such a structure and two functions set8mask and set6mask initialise masks using a convenient syntax. For example, a mask for line chewing can be described by

```
masktype end;

set8mask( x  1  x
          0  1  0
          0  0  0, &end);
```

and then applied to an image to delete line ends by

```
temp $= im;
for (i = 1; i <= 8; i ++) {
    aands(temp, thin(im, end));
    arotmask(&end, 1);              /* rotate mask once */
}
im $= temp;
dump(temp);
```

Tessellation

The binary masks, as well as giving the mask itself, also indicate whether hexagonal or square connectivity is needed according to which of set6mask and set8mask is used. Normally programs work extensively in one or other tessellation mode so that, except for the special case of binary masks, a global control is more useful than attaching a mode definition to each relevant function call or direction list. Therefore a **hex** flag is globally available which, when set, indicates hexagonal tessellation is to be used, rather than the default of square tessellation.

3.4 Subroutines

Clearly, to be of general use, IPC must allow a user to define his own image-handling functions to augment the built-in library functions. Whilst such support is provided, a major failure of IPC (in retrospect) is that it provides insufficient cosmetic help for such coding. Functionally, local images, the ability to pass images as arguments and to return images as the results of a function are all well catered for. However, the initialisation and deallocation of local images, the allocation of a global image as a return value and the consuming of dead temporary images must all be explicitly performed by the user. This means that 2 or 3 lines of code must be specified at the start and end of each function definition to perform such housekeeping. The code is a standard incantation and quickly becomes automatic but is a major source of programming errors and is conceptually difficult for new users. A future release of IPC should dispose of these problems.

A partial analogy can be drawn between images in IPC and string handling in C. To return a string from a function, global data storage must be allocated for the return value

and, before use, local strings must be initialised. Users
find string handling in C quite effective, but conceptually
difficult at first.

3.5 Access to CLIP4

Apart from the higher-level functions, three levels of
direct (low-level) access to CLIP4 are provided. Firstly,
all CLIP4 array instructions are available as standard
function calls. In particular, the basic cellular array
operator group of CAP4, namely

```
    SET set-def        ; define array instruction
    LDA a              ; load A-register from D-level a
    LDB b              ; load B-register from D-level b
    PST p              ; execute operation, put result in p
```

is available under IPC in the form

$$clipi(ima, imb, imp; set-defn);$$

where ima, imb and imp are single-bit images to take the
place of a, b and p in the corresponding CAP4 code. The
disc paging of images and global tessellation control of IPC
still applies to this packaged instruction.

Secondly, user-written CAP4 subroutines can be linked
into IPC programs in the same way as the library functions
are. The use of CAP is needed to attain full-speed binary
processing from CLIP4, the overheads imposed by IPC being
discussed below.

Finally, the program memory of the CLIP4 controller
(which includes a buffer for direct D-level access) is
available as a standard UNIX file. This provides a route
by which bit-planes can be moved between the array D-level
storage and the host computer's memory (for serial access).
It is used by the disc access primitives.

3.6 The Implementation of IPC

There are four basic elements to the IPC implementation.

1. A preprocessor to transform IPC source code into
    syntactically correct C code.

2. A mechanism for controlling CLIP4 from a C pro-
    gram.

3. A library of functions for storage allocation
    and other IPC internal operations.

4. A library of functions for actually processing
    images.

The limitations of the CLIP4 controller's instruction set
make it unfeasible to target a language like C directly into
the controller. Short of rebuilding the CLIP4 controller
(the only long-term solution), the best way to implement IPC
was thus to target programs into the host computer and perform all image processing functions by passing messages
between the host and CLIP4. This is a cumbersome arrangement and is the root cause of many of the limitations of
IPC. In particular, since the time taken to pass a message
from a user's C program to the CLIP4 controller is much
longer that a CLIP4 instruction cycle, each message must
correspond to many array operations to make full use of
CLIP4. To achieve this, a package of CAP4 subroutines is
loaded into the CLIP4 controller at the start of program
execution and IPC functions in the host then call these
low-level subroutines, passing to them relevant parameters
defining the locations of image data. As well as making
the use of single array instructions and binary image processing functions from IPC quite inefficient, this organisation means that IPC images must be compatible with CAP4.
Thus images have to be stored as contiguous bit-stacks and
no more than 32 D-levels can be processed by any particular
low-level CLIP4 function at once (CAP4 subroutines do not
use disc paging).

Having decided that IPC programs should be host
resident, an unmodified C compiler can be used with all
extended operations being carried out by library functions,
with the aid of a preprocessor to allow a more convenient
syntax for special operations.

4. THE MENU PROGRAM

Although in no way unique to the CLIP4 system, one of
its most heavily used facilities is the Menu program. This
program provides an image processing environment which is
easily operated by users with little or no programming
experience, a wide range of image processing functions being

available by typing one or two characters at the operator's keyboard. Images can be accessed from a television camera (by typing 'ip' for 'input') whereupon the image is immediately displayed. As each function is performed, the input image is replaced in the display by the result image. At any stage, images can be named and stored or recalled from store. An important feature of the system is that the Menu program is designed to remove from the user the need to consider the image length. Thus, for example, if two 6-bit images are multiplied together, the resulting 12-bit image is formed and used for subsequent calculations, but only the most significant 6 bits are displayed. In general, many of the commonly required image operations are supplied (arithmetic operations, filters based on 3 x 3 kernel convolutions, thresholding and histogramming, skeletonizing, ranking filters, and so on); the program prompts the user whenever a decision or parameter is needed.

This program, in association with a command system and system software for image handling (file manipulation, hard copy production, etc.), is restricted in capability by the limited space available for it in the PDP-11 host computer. Nevertheless, it has proved invaluable both to naive users who are unacquainted with CAP4 or IPC, and also to more experienced programmers who wish to use a fast interactive system for program development or for preliminary assessment of image data.

CHAPTER   FOUR

SERIAL   SECTION   RECONSTRUCTION

It is not difficult to propose designs for image processors embodying novel architectures. It is rather more difficult to build them. However, the ultimate test of a system is to apply it to real image processing problems, 'real' in the sense that they have not been invented by the processor designers with a view to demonstrating the better points of their designs.

In this and the following two Chapters, the three medically-related projects described were suggested by our collaborators and were therefore seen as excellent tests of CLIP4's capabilities.

The first project was described by Horace Ip in his thesis 'Automatic Detection and Reconstruction of Three-dimensional Objects Using a Cellular Array Processor'. It was carried out in collaboration with St Bartholomew's Hospital Medical College in London.

1. INTRODUCTION

Reconstruction from a set of serial slices provides a means whereby the spatial relationship of the internal structures of a three-dimensional object can be extracted and studied quantitatively. This project concerns the three-dimensional morphology of the vascular structures of the carotid body, which is a highly vascularised organ with the highest blood flow rate of any tissue in the body [26].

The task of reconstructing the three-dimensional vascular structure contained inside a series of histological tissue sections can be broken down into the following computing processes:

1. Alignment of successive tissue sections.

2. Segmentation of the blood vessel cross-sections present on each section.

3. Compilation of the segmented blood vessel cross-sections into an internal computer model of the associated three-dimensional vascular structure (tree) suitable for display and morphological analysis.

This Chapter reports the design strategy and the algorithms which take advantage of the parallel processing power of CLIP4 in automating the above subtasks. It should be noted that the basic methodology of reconstructing three-dimensional objects from a series of two-dimensional slices is the same whether the slices are produced from a microtome or from Computer Aided Tomography (CAT) or a Magnetic Resonance (MR) scanner. The various algorithms developed for this work may be applicable to serial slices resulting from these and other forms of imaging modality.

2. MATERIAL AND APPROACHES

Serial sections of the left carotid body of a rat, 2 μm thick, were prepared and stained such that the cells in the section assume a purple colouration which provides a dark, textured background in the presence of the bright (perfused) vessel interior regions. Corresponding areas of successive sections were digitised into 96 x 96 pixel, 6-bit images with a pixel resolution of 1.3 μm (see Figure 4.1).

As the spatial resolution of the two-dimensional sections is comparable with the section thickness, the sequence of section images, when suitably aligned, constitutes a set of three-dimensional data. Moreover, with the inter-section distance (i.e. section thickness) small compared with the size of the objects of interest, the following assumptions concerning the appearance of consecutive section images can be made: (1) the numbers of objects found on adjacent images are similar, and (2) the position and size of corresponding objects in adjacent images are similar barring small distortions arising from the slice preparation process. These assumptions are used in the design of the algorithms for the reconstruction subtasks.

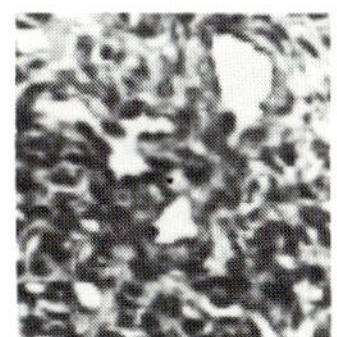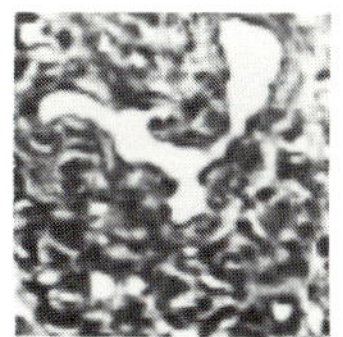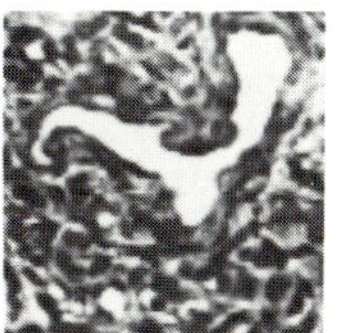

Figure 4.1   Three consecutive digitised tissue sections of the rat carotid body.

3. ALIGNMENT

The process of aligning successive sections is aimed at reconstructing the original spatial relationship of the blood vessel cross-sections in the third dimension, which is

lost in the slicing process.   Since it is not possible to
implant fiducial marks into the section without severely
distorting the objects inside, an alternative is to define
an alignment criterion based on parameters derived from the
aligning images.   The set difference of the major blood
vessel cross-sections extracted (by thresholding the image)
from the aligning images was chosen.

Given two binary images **B1** and **B2** the set difference
measure can be calculated in CLIP4 as follows:

$$n = |\textbf{B1} \; @ \; \textbf{B2}|$$

where the volume operator $||$ sums the grey values of all
the pixels in an image and @ denotes the exclusive OR operator.   In this case, the computation of n requires only two
basic instructions on CLIP4. The set difference measure, as
defined above, when minimised for the two images by
translating and rotating one with respect to the other,
gives alignment results similar to those obtained manually
based on the human experience of visual best fit [27].

Coarse alignment is achieved by evaluating the centre of
mass of the two thresholded images and shifting one image
with respect to the other such that the two centres of mass
coincide. The centre of mass of a binary image can be computed in CLIP4 by first ANDing the binary image with a ramp
image (in which each pixel grey-level is equal to its co-
ordinate) in the appropriate axis such that each white pixel
point is labelled with its co-ordinate. The volume operator
is applied to sum the total grey volume in the resultant
image and that of the binary image.   Dividing the former by
the latter yields the co-ordinate value of the centre of
mass.

Fine alignment proceeds iteratively using a hill-
climbing technique. At each stage, the values of the set
difference measure are evaluated, in predetermined increments, for four out of six possible linear transforms,
namely: vertical and horizontal shifts (in both directions)
and rotation (in both senses).

The control flow of the algorithm is designed to
minimise computations by performing image rotation only when
image shifts in all directions fail to further reduce the
set difference of the aligning images. The same sequence of
transformations is then applied to align the grey section
images. An iterative technique is chosen because, when
using CLIP4, an image can be shifted by one pixel in a single instruction.   Similarly, image rotation can be

implemented as a controlled sequence of image shifts in the appropriate directions [28] and does not involve matrix multiplications.   On average, a pair of section images can be aligned in 10-15 seconds.

4. BLOOD VESSEL SEGMENTATION

The alignment process produces a sequence of aligned section images which forms the input to the segmentation program.

4.1 Thresholding

The p-tile method [29] for threshold selection is usually applicable only when the total area occupied by the objects in the image is known in advance.   In the present case, a threshold selection algorithm, which is based on matching the p-tile values for objects found on adjacent section images, is defined as follows.

For the nth section image:

1. Calculate the percentage-area, k_{n-1}, occupied by blood vessel cross-sections in the (n-1)th image.

2. Search for a threshold value, t, which minimises $|k_n^t - k_{n-1}|$, where k_n^t is the percentage-area occupied by blood vessel cross-sections in the current image.

Algorithmically, the threshold selection algorithm can be expressed as follows:

$$k^t = 100 * |\mathbf{P}_t| / N$$

where t′ is chosen such that

$$|k_n^{t'} - k_{n-1}| = \text{Min}_t |k_n^t - k_{n-1}|$$

where $\mathbf{P}_t$ denotes the section image thresholded at t, t′ is the selected threshold for the current image and N is the total number of pixels in the image.   Thresholding in CLIP4 is implemented as a series of pointwise binary logic operations, and has complexity proportional to the bit-length of the image involved. In practice, only a few threshold values centred around the one chosen for the previous section image are tested.

In order to avoid segmentation errors propagating through the image sequence and to compensate for variations in stain density between sections, the threshold value is modified on the basis of current image grey-value statistics. This is accomplished by using the set of binary objects (produced by the above threshold selection method), which overlap with blood vessel regions already found on the previous section, as masks for pixel selection. Values of pixels within these masks are used as grey-value samples from which a modified threshold value is calculated. Figure 4.2 illustrates this thresholding method.

4.2 Two-dimensional Object Discrimination

A set of binary objects is obtained from the above thresholding process. As well as regions corresponding to blood vessels, it contains those arising from artefacts such as tissue tears and intercellular gaps. These objects are discriminated initially on the basis of a priori knowledge of blood vessel shape, size and boundary characteristics, and subsequently by using a criterion of continuity along the z axis, i.e. blood vessels are expected to be continuous through at least several consecutive slices.

Discrimination based on the extracted object feature values is implemented in parallel using feature images. The shapes of the objects in the feature image are identical to the corresponding ones in the original image. The value of every pixel belonging to each object is the same and is equal to the value of the object feature. A feature image is computed for each feature which is used in the discrimination process.

As an example of this process it is possible to discriminate in parallel a set of objects based on the ratio of area to perimeter squared. First, feature images for both the area and the perimeter of the objects are computed. The perimeter feature image is multiplied by itself and the area feature image is divided by the product image. (Both operations can be implemented using pointwise parallel arithmetic operations in CLIP4.) Discrimination can then be performed on the basis of thresholding the resultant

(a)   (b)

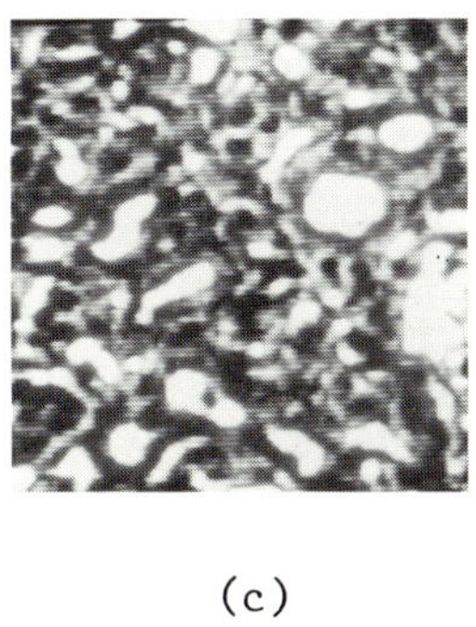

(c)

Figure 4.2   The threshold selection process for a tissue section image.   (a) Thresholded image (at grey level
30) using the p-tile matching method.   (b) The set of pixel
sample masks chosen from thresholded regions in (a) which
overlap with blood vessel regions in the previous section.
(c) The refined thresholded image (at grey level 28).

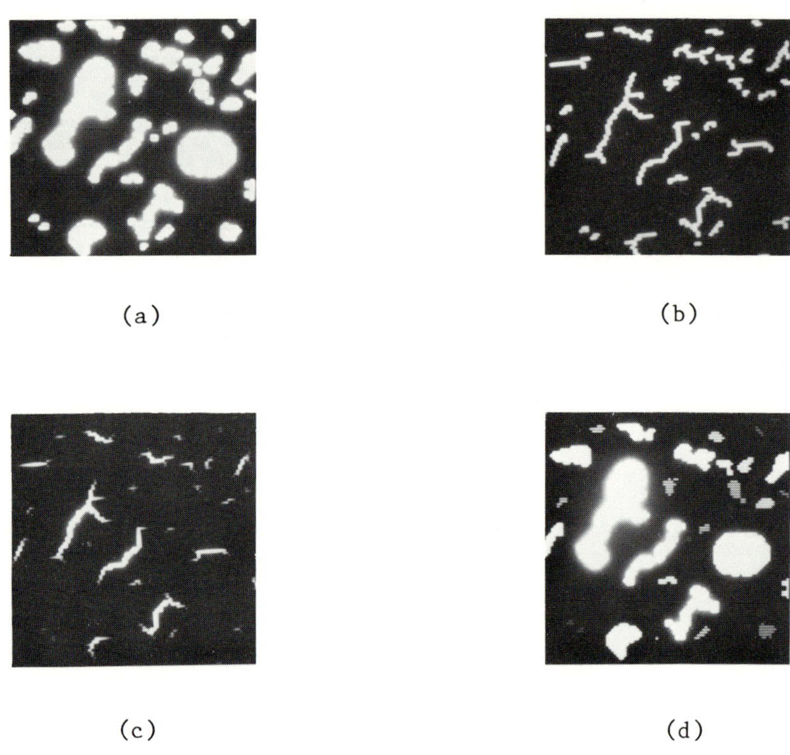

(a)   (b)

(c)   (d)

Figure 4.3   Construction of a feature image.   (a) The original image.   (b) The set of skeletons of the objects in (a).   (c) The result image from the skeleton length algorithm.   (d) The skeleton length feature image.

image at appropriate values. (Note that threshold selection algorithms can be used at this stage for discriminant analysis.) A serial computer would need to process and discriminate each object one at a time. Figure 4.3 illustrates the computation of a feature image based on the lengths of the medial axes of a set of binary objects.

The use of feature images is only advantageous for a parallel computer when a parallel algorithm is available for computing the feature image for the objects involved. Parallel algorithms have been developed for computing the feature images of the major and minor axes [30], the area, perimeter and the sum of grey values [31] for sets of objects.

The thresholded regions are classified into separate groups using the above technique in conjunction with a decision-tree (see Figure 4.4), which is designed from knowledge of the two-dimensional shape, size and the gradient characteristic of blood vessels in the section image. This process divides the objects into high-confidence regions, query regions and artefacts.   High-confidence regions are usually large objects with relatively smooth boundaries.   Query regions are medium-size objects which have failed a test designed to detect high-contrast points along the object boundaries.   These high-contrast points result from the grey-value difference between the bright vessel interiors and the darkly stained areas of the epithelial nuclei (nuclei of cells which form the blood vessel wall).   Artefacts are small and irregular regions.

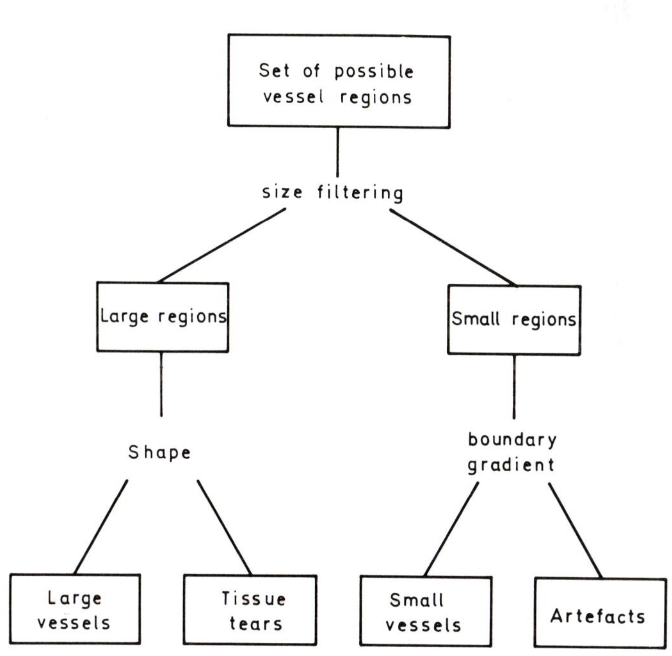

Figure 4.4   A decision-tree for blood vessel discrimination.

4.3 Three-dimensional Object Discrimination

Three-dimensional information is utilised in the discrimination process by constraining the blood vessel to be continuous through several consecutive sections and by matching features of prospective corresponding cross-sections.

Regions which have been found to have corresponding areas in the images above and below are assigned affirmatively as blood vessel cross-sections. The object labelling operation is used in this process to identify and extract corresponding regions in parallel. Similarly, the feature images can be used to implement feature matching so that correspondence of objects can be further substantiated or tested on the basis of their features.

Three-dimensional information can also be used to detect missing blood vessel cross-sections which have been squashed during the slice preparation process. Missing cross-sections can be interpolated and reconstructed from information on related cross-sections found in neighbouring sections. Figure 4.5(a) shows the output of this stage of the process for a sequence of images.

5. ISOLATING THE THREE-DIMENSIONAL STRUCTURE

The segmentation program outputs a sequence of binary images where white points denote regions of blood vessel interior. The sequence may contain cross-sections belonging to several vascular structures. An algorithm for isolating a three-dimensionally connected object has been developed to separate individual vascular structures.

5.1 Three-dimensional Connectivity

The 26-connected local neighbourhood of a volume element (voxel), $v(x,y,z)$, in a set of three-dimensional data can be defined as follows:

$$N_{26}(v_i) = \{ \; v_j \; | \; d_{26}(v_i,v_j) = 1 \; \}$$

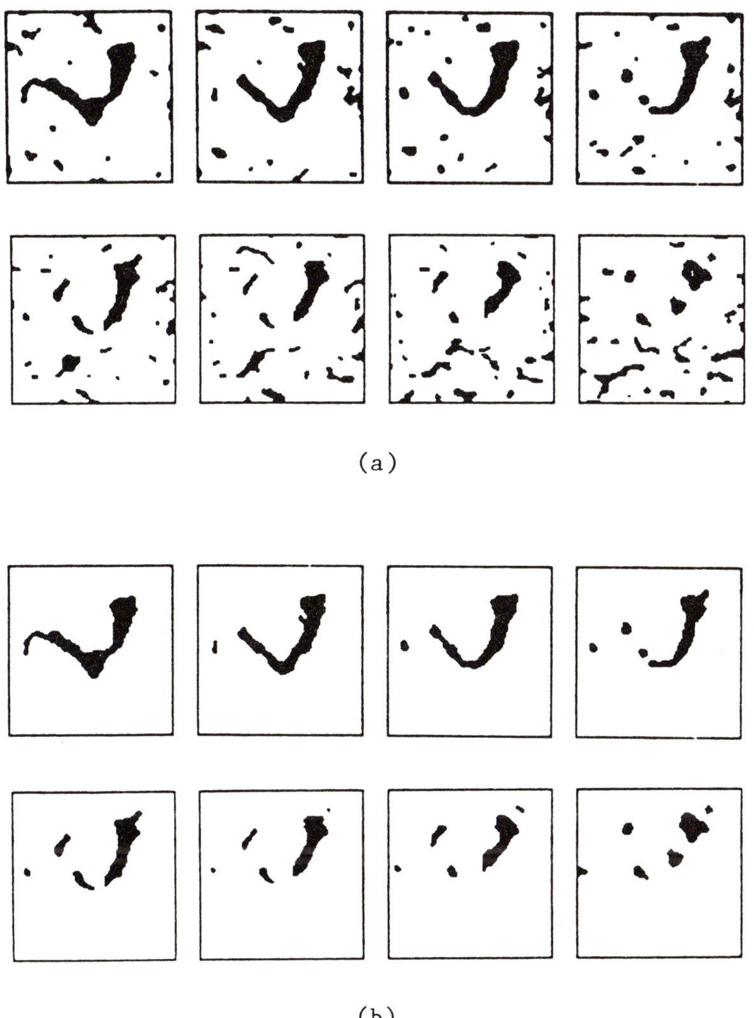

(a)

(b)

Figure 4.5   (a) A sequence of segmented blood vessel
images.   (b) The cross-sections of a single vascular structure isolated from (a).   (In this Figure the blood vessels
are shown in black.)

where

$$d_{26}(v_1,v_2) = \text{Max}(|x_1 - x_2| , |y_1 - y_2| , |z_1 - z_2|)$$

and (x_1,y_1,z_1) and (x_2,y_2,z_2) represent the voxels v_1 and v_2 respectively.

Since a plane of voxels is represented by an image in the image sequence, the 26-connected components in two adjacent images, say **Bn** and **Bn-1**, can be detected by expanding **Bn-1** once and using the transformed image to label the other image in an object labelling operation. The resulting image contains cross-sections in **Bn** which are three-dimensionally connected to those in **Bn-1**. Therefore, given one point which is a member of the object volume, a three-dimensional vascular structure can be isolated by successively labelling the images in conjunction with a controlling search strategy which guarantees total isolation.

Here, the processing power of an array computer can reduce the complexity of the isolation process from O(N), where N is the number of voxels constituting the three-dimensional object, which is typical if the isolation process were computed in a serial machine, to O(L) where L is the number of images containing the same object.

In order to make effective use of the fast local memories in CLIP4, the isolation algorithm uses a data partitioning scheme to divide the image sequence into blocks of consecutive images. The number of images inside each block is chosen such that they fit, together with the output and working images, inside the fast memories of CLIP4. Partial objects inside each block are computed independently first, then jointly through a control flow structure which decides which block needs to be re-entered (usually due to the presence of recurrent blood vessel branches). The output of this process is a sequence of binary images containing only one three-dimensional object (Figure 4.5(b)).

6. INTERNAL REPRESENTATION OF VASCULAR STRUCTURE

Although the image sequence representation of the reconstructed vascular structure can readily be used in analysis and display in CLIP4, a more compact representation is needed if the structure is to be studied using a serial computer. For this purpose a connected graph representation

of the vascular structure is built from the sequence of
binary images containing the structure.   Each node of the
graph represents a blood vessel cross-section and the
correspondence between cross-sections in adjacent sections
is represented by the arcs joining corresponding nodes in
the graph.   Within each node, a set of features such as the
position, area and boundary codes of the associated blood
vessel cross-section is stored.   Figure 4.6 illustrates
this representational scheme.

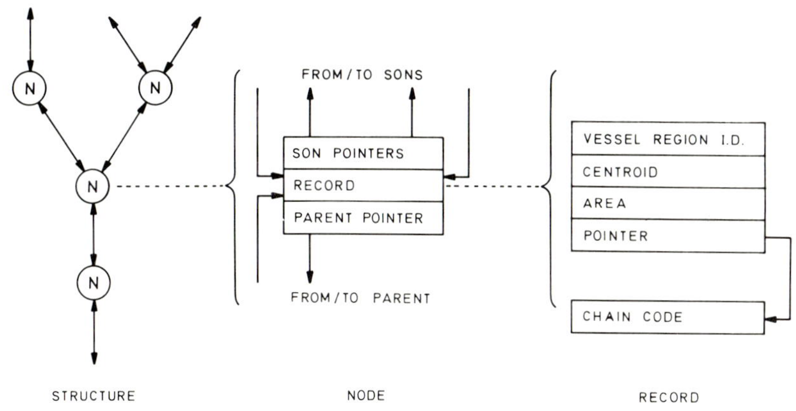

Figure 4.6   The representational scheme for a three-dimensional vasculature.

Since each cross-section is extracted and has its
feature values determined serially, the conversion time is
dependent upon the number of cross-sections representing the
three-dimensional object.   This highlights the need for an
efficient interface for converting between the basic data
structures of CLIP4 and those of the host machine.   The
advantage of using graph representation is that conventional
techniques of computer graphics and graph traversal algorithms can be applied to display and analyse quantitatively
the three-dimensional properties of the vascular structure.

7. DISPLAY AND MORPHOLOGICAL ANALYSIS

Currently, there are three forms of display for the visualisation of the reconstructed vascular structure:

1. The medial axis (skeleton) of the vasculature is displayed with lines joining connecting cross-sections in consecutive sections. The lines are colour-coded according to the diameter of the associated portion of the blood vessel. Such displays facilitate the studies of branching patterns without the hindrance of the surfaces which are present in a realistic display.

2. Successive cross-sections of the vasculature are stacked up in the correct order with hidden lines removed, producing a realistic view of the object.

3. A stereo-pair of the vascular structure is displayed, with one in red and the other in green. Special viewing glasses can then be used to perceive the relationships of the various vascular branches in space.

The perspective displays can be shown for different viewpoints. The graph representation of three-dimensional vascular structures can easily be converted to inputs suitable for three-dimensional surface tiling algorithms [e.g. 32, 33] and, subsequently, three-dimensional shaded display of the vasculature.

Graph traversal algorithms (either breadth-first or depth-first search) can be applied to traverse the graph representation of the vasculature, collecting and compiling appropriate morphological data, such as branching order, blood vessel branch length and size, etc. for analysis.

8. CONCLUSIONS

The use of CLIP4 in a highly data-intensive computation task has been described. This project provided an opportunity to use CLIP4 in a variety of image processing tasks, namely image registration (alignment), object segmentation and processing of three-dimensional data by extending the notion of two-dimensional connectivity and local neighbourhood to three dimensions.   Details of this project and the implementation of the reconstruction system can be found in [30].

For this application, the criterion for determining correspondence is based on connectivity within a three-dimensional local neighbourhood.   In general, the criterion for object correspondence can be expressed as a transformation based on affine transforms and/or morphological operators.   After the initial transformation, an object labelling operation can be used to detect and extract corresponding objects.   For example, in the case of a temporal image sequence in which corresponding objects in succeeding frames differ in displacement due to motion, methods described in this Chapter can be adapted to detect and extract corresponding objects and, subsequently, to track moving objects in the sequence.

More particularly, some of the techniques described above can be applied to the reconstruction of three-dimensional objects found in a sequence of images generated from non-invading imaging modalities such as CAT or MR. The system has been used successfully, after minor modification of the parameters in the segmentation module, to reconstruct phantom objects from a series of positron emission images and to detect a three-dimensional rock layer from a series of echograms used in seismological studies.

CHAPTER FIVE

COLONY COUNTING AND ANALYSIS

Despite the enormous amount of effort now being invested in research into automatic image analysis techniques, it is rare that one finds such methods in routine use. Disappointingly, many of the techniques which are developed are not sufficiently robust to be trusted to work on real data, much of which seems to exhibit properties which never appear in the test data used to develop the analysis algorithms.

The following Chapter describes a process which, perhaps because it uses somewhat simple image processing operations, has proved to be sufficiently reliable for it to be employed in a routine way by collaborating researchers inexperienced in image analysis. This work formed a part of the doctoral research programme of David Potter and is described in his thesis 'Analysis of Images Containing Blob-like Structures Using an Array Processor'.

1. INTRODUCTION

The initial task was to develop an automated system for counting and measuring the diameters of cell colonies, grown in small culture dishes, to aid the experimental work of Dr M Rosendaal and his colleagues in the Department of Anatomy and Embryology, University College London. The analysis of these colonies was routinely done by eye using a microscope, a very mundane job, taking on average 3 hours for 50 dishes. Before discussing the image processing involved, it is worthwhile describing the general background of the problem, the analysis of bone marrow and, in particular, the blood-forming system [34].

Bone marrow is a somewhat random mixture of cells of numerous sizes and morphologies, which make up the haemopoietic family of cells. These are the cells which produce the cells found in the blood such as macrophages and red blood cells.

It is widely recognised that haemopoietic tissue has a basic three-tiered arrangement (see Figure 5.1). First, there is the fundamental pluripotent stem cell population, which gives rise to most, if not all, of the cell lines found in haemopoietic tissue. These cells are few in number (0.2% of the total cells in the marrow), are morphologically unidentifiable, and are capable of self-renewal. Second, there are the committed precursor cells which are the daughters of differentiation events and commit the stem cells to one of several specific lines of cell development. They proliferate rapidly and expand their population by several division cycles. They remain morphologically unrecognisable and the population is small. The bulk of the marrow falls into the third category of maturing cells. These are morphologically recognisable cells and are the largest group because they continue, through a further series of amplifying divisions as they mature, to become functional cells which leave the marrow to populate the circulating blood and tissues.

Under the conditions of growth in the routine culture assays, the result is normally a mixture of colonies containing granulocyte (G) and macrophage (Mϕ) cells, important

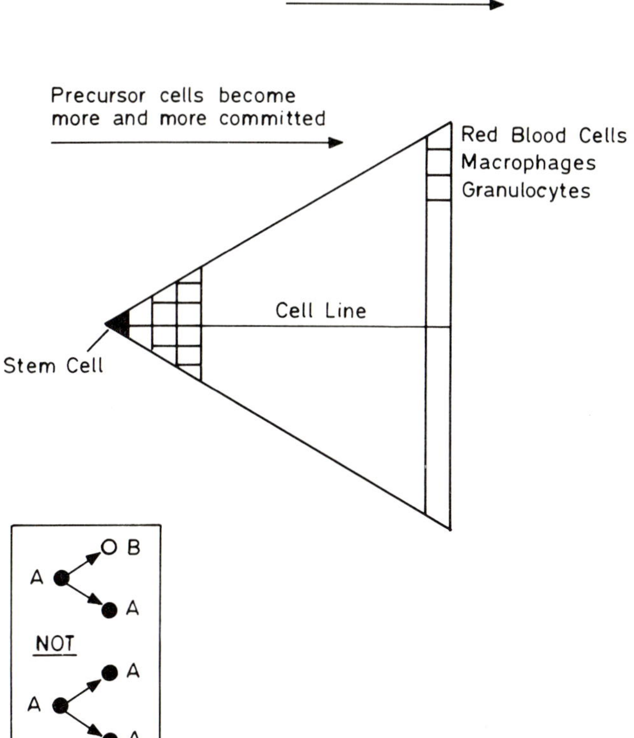

Figure 5.1   The haemopoietic system.   In the boxed illustration, a stem cell A produces a precursor cell B (A → A + B).   It does not simply divide (A → A + A).

in the auto-immune system. One or more of a family of stimulatory factors is required to initiate and promote the growth and differentiation of these colonies.

Mixed colonies of G and Mϕ are formed by certain cells in normal bone marrow (NBM) when a source of granulocyte macrophage colony stimulating activity (GMCSA) is incorporated in the culture medium. It has been found that both G and Mϕ arise from a single kind of committed precursor, GMCFC, or granulocyte macrophage colony-forming cell.

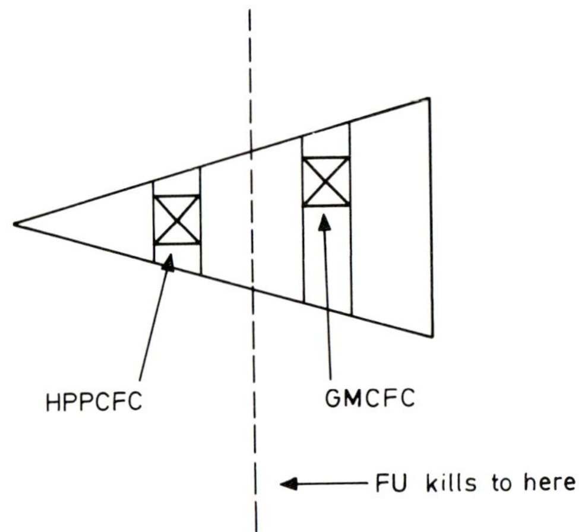

*Figure 5.2   Effect of fluorouracil (FU) on the haemo-
poietic system.   FU kills all cells to the right of the
dotted line.*

Cells in mouse marrow have also been detected which will only form colonies in culture when a combination of stimulatory factors are used – GMCSA and synergistic activity (SA). These cells, denoted high proliferation potential colony forming cells (HPPCFC) are difficult to culture when there are large numbers of GMCFC present. Therefore, donor mice are usually treated with the cytotoxic drug, 5-Fluorouracil (FU) with the subsequent deletion of most of the GMCFC (Figure 5.2). It has been found that HPPCFC are the precursors of GMCFC.

2. EXPERIMENTAL METHOD

The experimental methods for the cultures will be briefly described. There are four main stages in the experiment:

1. The mouse is injected with the cytotoxic drug, FU, to kill the cycling cells.

2. After a certain period (normally two to ten days), the femoral bone marrow is collected. An animal's marrow that is allowed to recover for m days is denoted FUm.

3.   After placing the desired amounts of GMCSA and
     SA in the colony dish, the marrow cells are
     added.   They are left in optimum culture condi-
     tions for 11 days.

4.   The colonies are stained for aiding the analysis
     and then presented to CLIP4.

3. REQUIREMENTS OF THE IMAGE PROCESSING SYSTEM

Baines et al. [35] showed that the cellularity (number
of cells) in a colony, when measured after various times of
culture, was related to the colony diameter.   Table 5.1
shows this relation.

Table 5.1   Relation between cell diameter and cellularity

Colony diameter (mm)	Cell number (average)*	Doubling divisions†
0-0.2	50	7
0.2-0.5	400-1000	9
0.5-1.0	8000	13
1.0-1.5	11300	14
1.5-	50000-	16

* (approximate)

†The term doubling divisions refers to the number of
times the cell has divided, i.e. the number of cells equals
2^n where n is the doubling division.

Hence, knowledge of the area of each colony gives some
indication of the number of cells present.   Also, in the
particular experiments in progress, the area is an indication of whether the cells are HPPCFC (large colonies > 50000
cells) or GMCFC (small colonies 50-5000 cells).   The HPPCFC
cells are larger because they are considered to be developmentally younger in the dividing process.   The first
requirements of the system became:

1.   To count the total number of colonies in a dish.

2.   To sort the colonies into diameter classes.

It was not possible to use the five classes of diameter shown in the table because of the resolution of CLIP4. With only a 96 x 96 array and the dishes being 30 mm in diameter, the maximum resolution was approximately 0.3 mm per pixel. Therefore four classes of diameter, D, were used: $D \leq 0.5$ mm, 0.5 mm $< D \leq 1$ mm, 1 mm $< D \leq 1.5$ mm, and $D > 1.5$ mm.

It was also decided to produce the integrated optical density (sum of the pixel grey levels) of each colony to see if there was any relation with cell number.

The system had to be simple and user-friendly, so that experimenters untrained in computer operation could use it, and moderately fast to make it practical.

4. THE SEGMENTATION ALGORITHM

Obviously the most important part of the system to be developed was the segmentation algorithm.   Once the colonies had been picked out from the background, it was a relatively simple task to analyse each colony.

A typical digitised image of a cell culture dish is shown in Figure 5.3.   As can be seen, the cell colonies are clearly visible against the background (due to the staining process), so a low-level algorithm for segmentation could well solve the problem.   In all dishes, a dark ring was found to appear around the edge.   This is due to excess stimulant medium which, because of mixing in the dish, has built up round the edge, and also to the slanting edges of the dish which produce a shadow effect.   Some colonies overlap with this ring and as area and density measurements would be badly affected, it was decided to use a masking operation to ignore the ring and colonies on or touching it. This is justifiable for two reasons:

1. Results are used only to give indications of trends, and it can be assumed that a constant small percentage of random size colonies grow on the sides of the dish.

2. The colonies at the edge are in a different support climate to those in the rest of the dish (due to more stimulant being present).

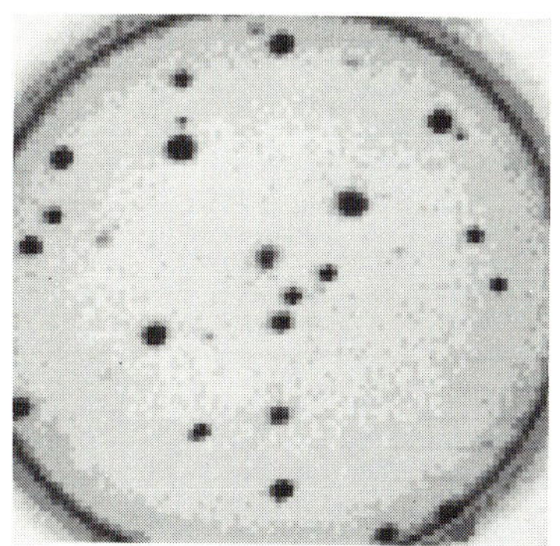

Figure 5.3   Digitised image of cell culture dish.

Using the Menu system available (see Chapter 3), global thresholding was applied to the digitised image to see if reasonable segmentation results appeared. The results were most encouraging and thus the main problem became the calculation of the correct global threshold value.

A global threshold is normally calculated from inspection of the grey-level histogram of the image. To improve the histogram prior to inspection, two preprocessing steps were employed: a correction routine to normalise the image to compensate for the uneven background and a simple filter to remove background noise. The correction program is a standard IPC subroutine package which takes four pictures of the blank background and then averages and scales them to a six-bit image. This image is used to normalise four input pictures of the dishes by division and truncation to force the resultant images also to six bits. The average of the four normalised images is the input image for the filter; the averaged background image is also used. The algorithm for the filter is as follows:

Algorithm 5.1   High-pass filter

Let F initially be the averaged background image, I the averaged normalised input image, and O the required filtered image.   WS1, WS2, and WS3 are temporary working space images.

1. F ← F - medr2(F)

2. F ← absvls(F)

3. max = mgl(F)

4. count = 0

5. WS1 ← I

6. WS2 ← expand(WS1)

7. WS3 ← WS2 - WS1

8. WS1 ← WS2

9. count = count + 1

10. If mgl(WS3) > (1.5 * max) go to (6)

11. For i = 1 to count in increments of 1 do step (12)

12.   WS1 ← shrink(WS1)

13. O ← WS1 - I

medr2, absvls, expand and shrink are IPC subroutines, specifying a median filter of radius 2, and pixels setting themselves to their absolute value, maximum or minimum grey level in their 3 x 3 neighbourhoods respectively.   mgl is a routine to find the maximum grey level in an image and is described in Algorithm 5.2.

The first three steps estimate the grey-level maximum amplitude of the high-frequency noise in the background image.   Then repeated expands of the input image are performed until the difference between subsequent images is less than 1.5 times the maximum noise level, indicating that colonies have been removed from the image.   The same number of shrinks is applied in case the background is uneven, producing an image representing an estimate of the background. The input image is subtracted from the background, hence inverting the output image so that colonies have higher grey levels than the background.

Algorithm 5.2   Find the maximum grey level (mgl) in an image

Let **I** be the input image.

1. Set **WS2** to 1

2. For i = length(**I**) to 0 in decrements of 1 do steps (3) to (6)

3.   **WS1** ← plane(**I**,i)

4.   If volume(**WS1**) = 0 go to (2)

5.   **WS2** ← ands(**WS2**,**WS1**)

6.   If volume(**WS2**) ≠ 0, **WS3** ← **WS2** otherwise
        WS2 ← **WS3**

7. mgl = volume(ands(**I**,firstpoint(**WS3**)))

plane returns the ith plane from the top of image **I** and length returns the number of planes in the specified image. Basically, the algorithm is inspecting bit-plane by bit-plane, starting with the most significant, and keeping track of potential maximum grey levels in image **WS3**. The grey-level histogram of a typical output image from these two preprocessing steps is shown in Fig. 5.4.

The algorithm chosen to determine the threshold value was that described by Ridler and Calvard [36]. It is an iterative procedure whereby the threshold t is updated at

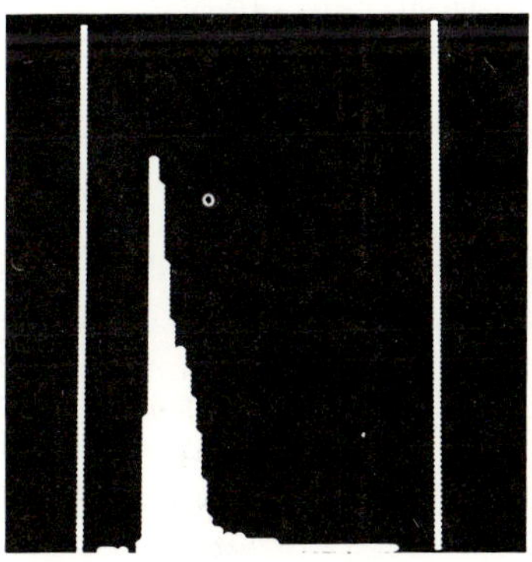

Figure 5.4   Histogram of the image in Figure 5.3 after background normalisation and high-pass filter.

each iteration. The first estimate, t_o, is simply the mean grey level of the entire image. The nth estimate, t_n, is given by:

$$t_n = \frac{\text{mean}(\textbf{in} > t_{n-1}) + \text{mean}(\textbf{in} < t_{n-1})}{2}$$

where $\text{mean}(\textbf{in} > t_{n-1})$ is the mean level of grey values of image **in** greater than t_{n-1} and $\text{mean}(\textbf{in} < t_{n-1})$ is the mean level of grey values of image **in** less than t_{n-1}. As this algorithm is very efficient on a parallel machine it has been included in the IPC subroutine library. The algorithm produces good results for a wide range of images and picks out colonies well.

To remove the edge effects a mask was generated by approximating an image to a circle function using the relation:

$$r^2 = x^2 + y^2$$

Algorithm 5.3   Generate a circle mask, radius N

1. Generate a ramp in the x direction, 7 bits in length (i.e. label each pixel with its x co-ordinate)

2. Subtract 47 from the ramp to obtain the centre in middle of the image

3. Square the resultant image

4. Repeat steps (1) through (3) for the y direction

5. Add the x and y images together

6. Square-root the image

7. Threshold the result at N to give a circular mask of radius N

In practice, the mask had a radius of 47 pixels to gain the most resolution. Figure 5.5 shows the result of segmenting the image in Figure 5.3 after masking to remove edge effects.

The next step was to calculate each colony's area, A, and integrated optical density and to relate the areas to corresponding diameters (see Table 5.2).

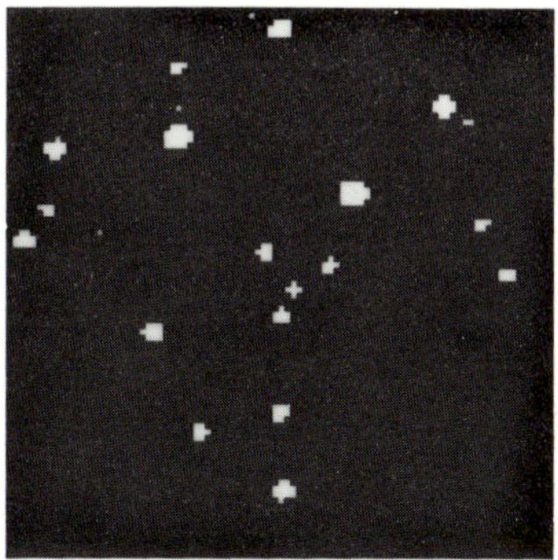

Figure 5.5   Segmented result of the image in Figure 5.3.

Table 5.2   Relationship between colony diameter and area in pixels

Colony diameter (mm)	Area (pixels)
D $\leq$ 0.5	A $\leq$ 2
0.5 < D $\leq$ 1.0	2 < A $\leq$ 8
1.0 < D $\leq$ 1.5	8 < A $\leq$ 18
D > 1.5	A > 18

To determine the area and density of each colony, further IPC routines were used: ands and andnots which are logical functions, volume(**im**) which returns the sum of grey levels in image **im**, firstpoint(**im**) which returns the top left point of the image, and label(**i1,i2**) which uses the second image **i2** to label the first image **i1**.

Algorithm 5.4   Determine individual colony parameters

Let **I** be the input normalised picture of the colonies, and **S** the segmented binary image of the colonies.

1. **WS** ← firstpoint(**S**)

2. **WS** ← label(**S**,**WS**)

3. area = volume(**WS**)

4. density = volume(ands(**I**,**WS**))

5. **S** ← andnots(**S**,**WS**)

6. If volume(**S**) > 0 go to (1) else stop

5. ASSESSING THE SEGMENTATION AND ANALYSIS ALGORITHMS

To judge the success of the algorithms, two important assessments of the results obtained need to be made:

1. Comparison of the results with those obtained by eye.

2. Measurement of the self-consistency of the algorithms, in order to comment on the reproducibility of the results (especially those due to redigitisation).

One hundred dishes were analysed both by using CLIP4 and by eye. The colony diameters were manually measured by Miss J Adam of the Department of Anatomy and Embryology at University College London. In counting the colonies, those at the edges which would have been masked out in the computer algorithm were ignored. The results obtained for the three largest diameter classes and for their combined total are displayed in scatterplot diagrams in Figure 5.6. The lowest diameter class (that ranging from 0 to 0.5 mm) had a very wide scatter biased towards the eye, indicating that the resolution was not sufficient to obtain accurate results. Each count recorded is the number of unmasked colonies of a particular diameter class in a dish. Points lying on the line y = x indicate perfect agreement; lines corresponding to a scatter of ±5% have been drawn. As more than one result can be plotted at the same point, the following legend was used: the numbers 1 to 9, upper case letters A to Z (10 to 35) and lower case letters a to z (36 to 61).

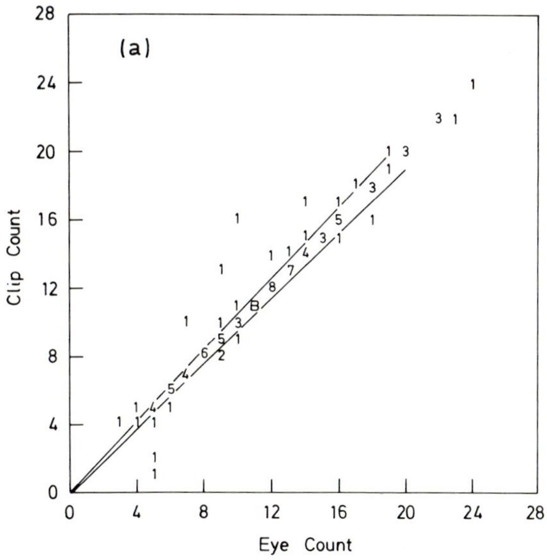

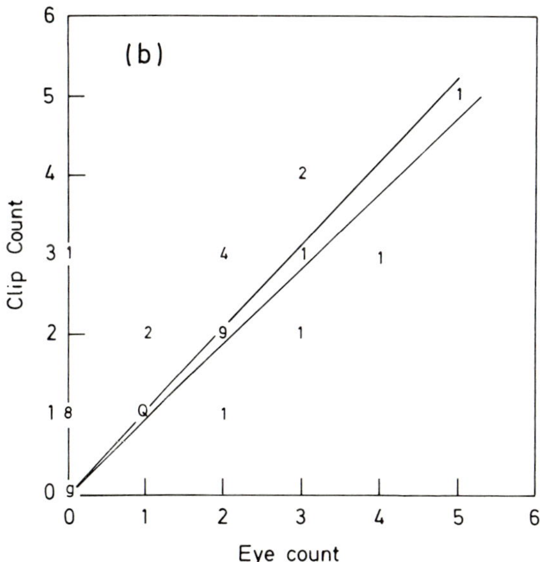

Figure 5.6   Scattergrams of computer measurement against eye measurement for various colony diameter classes: (a) 0.5 mm < D ≤ 1.0 mm; (b) 1.0 mm < D ≤ 1.5 mm.

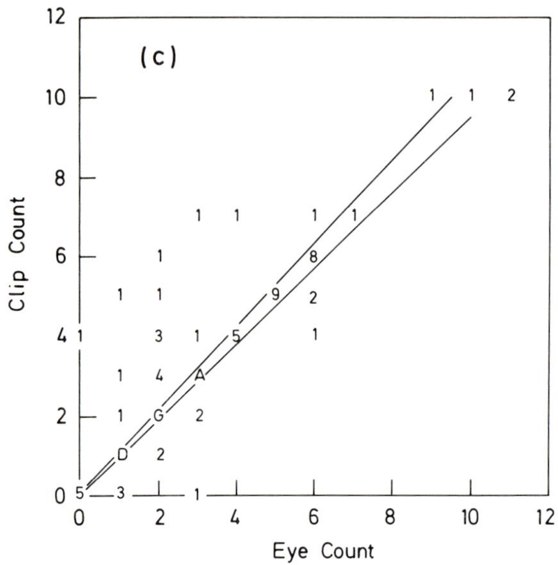

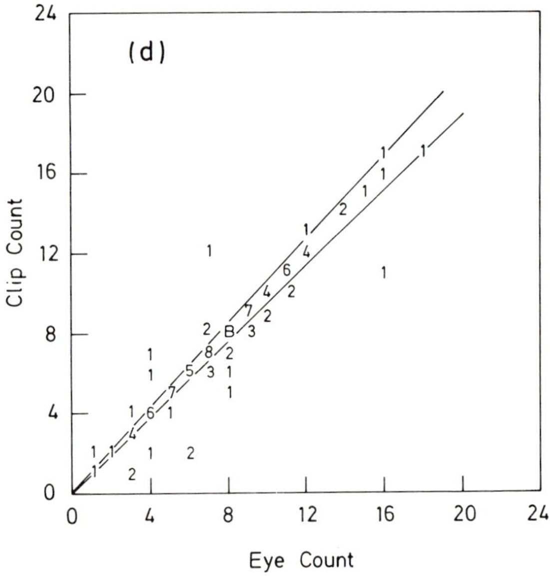

Figure 5.6 (continued): (c) D > 1.5 mm; (d) combined total of all classes.

In summary, the percentages of dishes within 5% error on counting for the different diameters and for the combined total were:

$$0.5 \text{ mm} < D \leq 1.0 \text{ mm} \qquad 70\%$$
$$1.0 \text{ mm} < D \leq 1.5 \text{ mm} \qquad 68\%$$
$$D > 1.5 \text{ mm} \qquad 79\%$$
$$\text{Combined total} \qquad 80\%$$

Another way of displaying the results is by plotting the percentage of the total number of dishes against the difference between the eye count and the CLIP count, d. Ideally, the peak should be at $d = 0$ with a normal distribution spread. Differences in spread to one side or another of the vertical axis indicate a bias towards certain types of error. The same four classes as in the scattergrams are plotted in Figure 5.7. Table 5.3 lists the observations found.

Table 5.3   Percentage of dishes with a given d for
             different diameter classes

Diameter (mm)	Eye count − CLIP count		
	$d = 0$	$-1 \leq d \leq 1$	$-2 \leq d \leq 2$
$0.5 < D \leq 1.0$	68	88	93
$1.0 < D \leq 1.5$	68	87	93
$D > 1.5$	79	99	99
Combined total	76	92	94

Negative values of d correspond to objects registered by CLIP4 but not by the observer. There are three possible reasons:

1. Error by the observer – the colony was not seen.

2. Noise or contamination marked as a colony by CLIP4.

3. Wrong area classification range (either by eye or CLIP4).

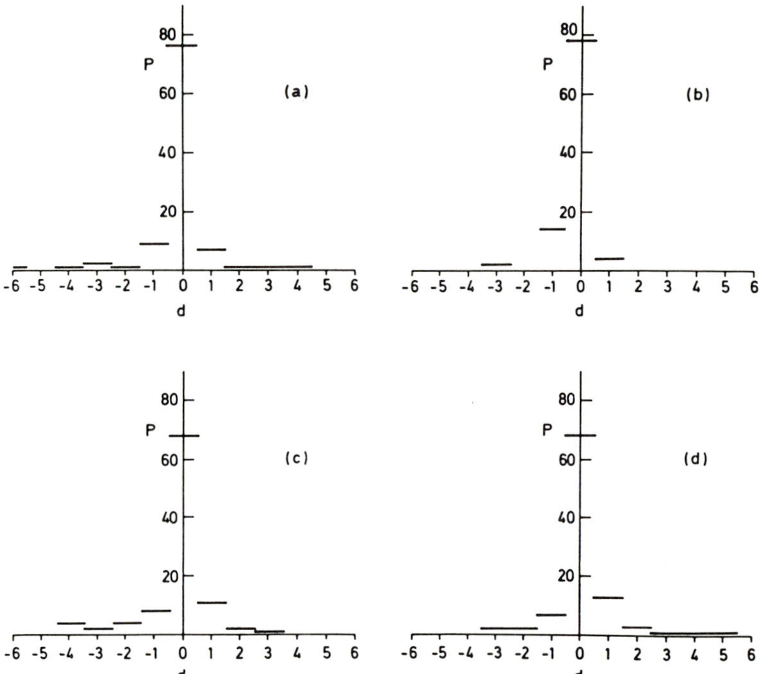

*Figure 5.7   Percentage of total number of counts (P)
against d, where d = (EYE COUNT - CLIP COUNT), for various
colony diameter classes: (a) 0.5 mm < D ≤ 1.0 mm;
(b) 1.0 mm < D ≤ 1.5 mm; (c) D > 1.5 mm; (d) combined
total of all classes.*

Positive values of d correspond to colonies seen by the
observer but not by CLIP4.   Reasons could be:

1. Error by the observer - non-existent colonies
    counted.

2. Wrong area classification range.

No bias can be seen towards positive or negative values
of d, which could have indicated a specific type of error.
There is an acceptable error by the observer.   A set of
batches can take up to six hours to analyse and errors in
counting must occur.

The largest error appears to be area misclassification.
Various experiments can be done to analyse the area misclassification errors in CLIP4.   The main reason for the

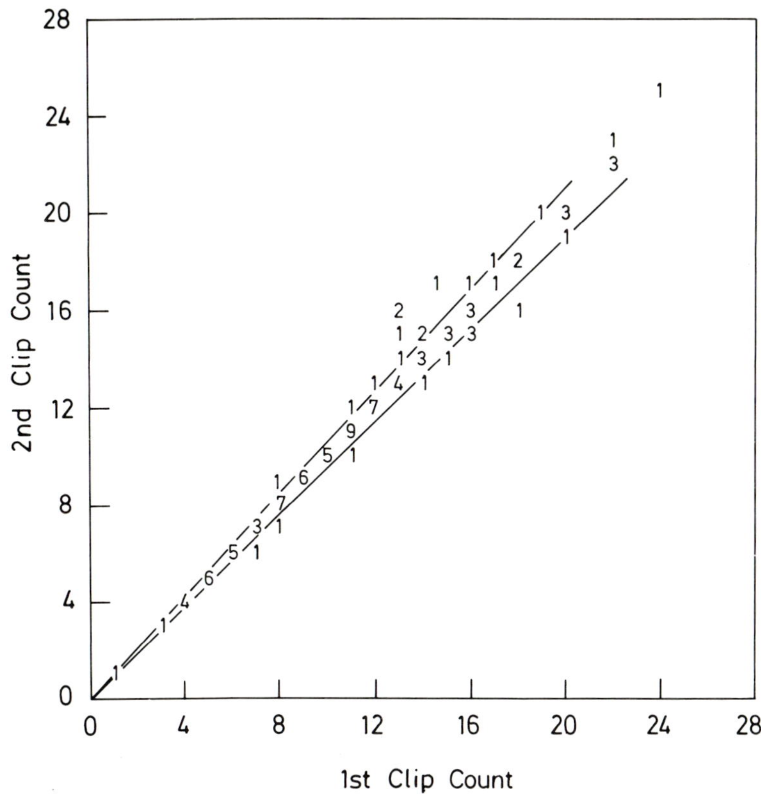

Figure 5.8   Effects of redigitisation on total colony count.

errors will be digitisation. Figure 5.8 shows a plot of one CLIP count against another, where the count is the combined total of colonies in the three top area classes in the dish. As can be seen the errors are very small, 79% having less than 5% error. This indicates that the overall segmentation process produces a consistent number of detected colonies when their diameter is greater than 0.5 mm. However, for the individual diameter classes the percentages within 5% error on redigitisation are:

$$0.5\text{mm} < D \leq 1.0\text{mm} \qquad 69\%$$
$$1.0\text{mm} < D \leq 1.5\text{mm} \qquad 64\%$$
$$D > 1.5\text{mm} \qquad 94\%$$

indicating that some colonies are classified into different diameter classes when the image is redigitised.

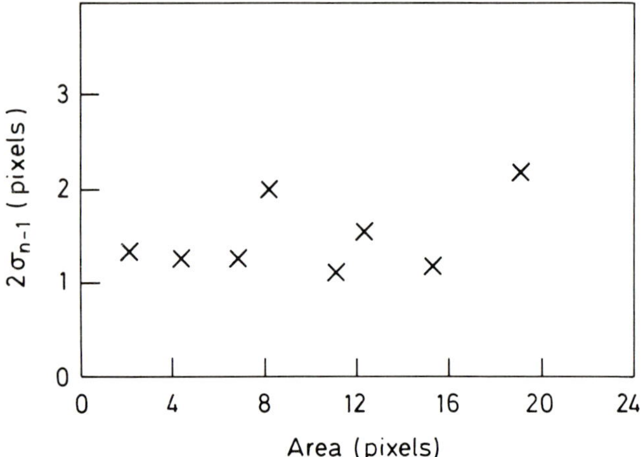

Figure 5.9   Error associated with area measurement of a colony.

As the distribution of colony sizes in each dish varies according to experiment, no misclassification error can be attached to each size range. Instead, an error can be applied to individual area sizes, calculated by measuring the spread of area measurements of various colonies through redigitisation of the picture. Figure 5.9 shows a graph of $2\sigma_{n-1}$ against the mean area of the colony where σ_{n-1} is the standard deviation of the area spread in pixels for a given colony measured over ten digitisations of the image. From this graph estimates of error can be attached to the area of a specific dish. The digitisation error can adequately account for the majority of errors in the counting of colonies by machine.

If the colony distribution is uniform throughout the dish, one can estimate the number of colonies lost through masking. For a set of 46 dishes, counts were made of the number of colonies totally within circles of various radii, the circles being concentric with the dish. The counts were converted into percentage of total count for that dish (i.e. count at radius = 47 pixels) and averaged for each radius. If the distribution were uniform, then the percentage count, C, would be proportional to the area of the circle. A plot of log C against log r, where r is the radius of the circle in pixels, should yield a straight line graph with gradient 2, i.e.

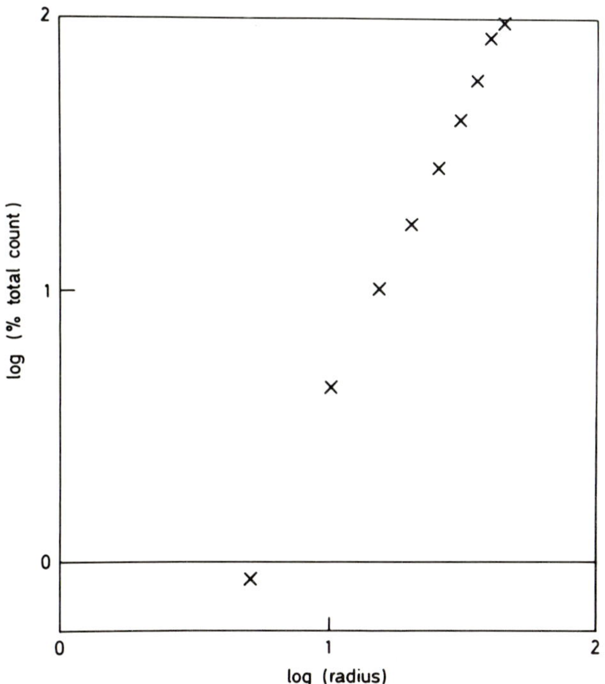

*Figure 5.10   Graph of log (percentage of total count)
against log (radius).*

$$C \propto area \rightarrow C = k\pi r^2$$

$$\log C = \log k\pi + 2 \log r$$

where k is a constant.

The graph is shown in Figure 5.10 and there is a very
good approximation to a straight line.   The gradient from a
least squares fit is 2.15 $\pm$ 0.03 which agrees well with the
uniform distribution reasoning.   Using the straight line
fit, the 100% count appears at r = 45 pixels, two pixels
less than the size of the mask, giving an estimate of the
effective dish size seen by CLIP4 when touching colonies are
ignored.   If the whole dish were to be included in the
analysis, the radius of the circle would have to be approximately 50 pixels, which would imply that about 23% more
colonies would be counted.

Two more possible errors must be considered.   As the
analysis of a dish takes approximately two seconds, the dish
might not have been long enough in the view of the camera to

remove the persistence of the old image thereby causing poor results. On comparing counts for each area category after ten seconds under the camera with those for normal conditions, no significant differences were found, the results being within the digitisation errors given previously.

The camera used has an automatic gain device, with no option of switching it off. Because of this feature, measurements on a single colony in a dish might be different from those on the same colony if it were one of many in a dish. To analyse possible errors, measurements were taken on a dish with only one colony. In steps, ink blobs were used to simulate extra colonies to see if there were significant changes in the measurements of the original colony. Again results were within the digitisation error.

6. TIMINGS

A typical dish is analysed, on average, in approximately 2.2 s of computer time. This can be broken down into the segmentation and analysis stages where the former requires 0.4 s but the analysis time is proportional to the number of colonies in a dish and can vary from 0.2 s to 4 s. In addition dishes have to be wiped dry to remove condensation and inspected for contamination. On average the total time taken for all the operations is 6 s per dish.

7. THE SYSTEM AS SEEN BY THE USER

The operator sees the final developed system as a menu of various commands. After instructions have been given on setting up the camera, light box and specially built holder for the dishes, the commands available in the menu are:

b — new batch to analyse

The computer asks for a batch name following which it gives a prompt indicating it is ready for the first dish. Once the dish has been positioned, the letter p on the keyboard is pressed and CLIP4 processes the image. On the monitor, bit-plane 4 of the output memory shows the circular mask and bit-plane 5 the segmented image. This is mixed with the television image of the dish so that the operator can see how well the segmentation algorithm has performed.   An audible signal indicates the end of processing and the 'ready' prompt returns.   When a particular batch is finished one can exit back to the main menu.

? – help command

This gives full instructions to assist a new user, or instructions on various sections as required.

l – list the batch names which have been analysed

p – print data

After listing the batches required (default is all batches), a hard copy is given of the relevant results.   The data is stored in a C language structure format making any information required readily accessible.   At present, a typical printout consists of (1) the number of dishes in the batch, (2) the mean number of colonies in each dish, (3) the mean area of the colonies in the batch, (4) the mean integrated optical density of the colonies in the batch, and (5) the number of colonies in each size class in the batch.

v – verbose flag

By setting this flag, full details of each colony in each dish can be given instead of only statistical data such as mean colony number, mean area, etc.

x – analyse data

On being given a batch name, the results (see p above) are displayed on the terminal.

A program has also been written for displaying the data as a histogram on the graphics terminal.   The number of colonies (for a batch or a set of batches) against any area range and step size can be plotted.

The system has proved to be successful and simple to use and over one hundred experiments have now been analysed using it.   Those described in the next two sections have been carried out jointly with Dr M Rosendaal and colleagues.

8. RELATIONSHIP BETWEEN GREY LEVEL AND CELLULARITY

An investigation was carried out to see if there was any relationship between the number of cells (cellularity) in a colony and a function of the grey levels representing that colony. Consider a pixel area of a colony with incident light intensity I_o being reduced to I_x after transmission. If the thickness of the colony section is x, one would expect that

$$I_x = I_o e^{-kx}$$

where k is a constant representing the optical density. The thickness, x, of the colony is proportional to the number of cells, p. Hence

$$I_x = I_o e^{-k'p}$$

The relationship between g, the observed grey level and I_x can be written as

$$g = mI_x + c$$

where m and c are constants derived from the parameters of the video system. By adjustment of the video amplifier in the CLIP4 hardware, and by studying the histogram of an opaque object in a colony-free dish, c can be approximately set to zero. Hence

$$\log g = -k'p + \log mI_o$$

If the number of cells in a section of colony can be determined, plots of the corresponding sum of the logarithms of the grey levels of constituent pixels against number of cells should produce a straight line.

The experiment consisted of taking a close-up picture of a colony such that it almost filled the 96 x 96 array. By calculating a threshold as described earlier, the constituent pixels of the colony were marked on the printout of the grey levels of the image. Using an electron microscope, serial sectioning of the colony was carried out, and the cells were then counted.

Figure 5.11 shows a graph of the sum of the logarithms of the grey levels forming the colony against the number of cells in the colony. A linear relationship is observed.

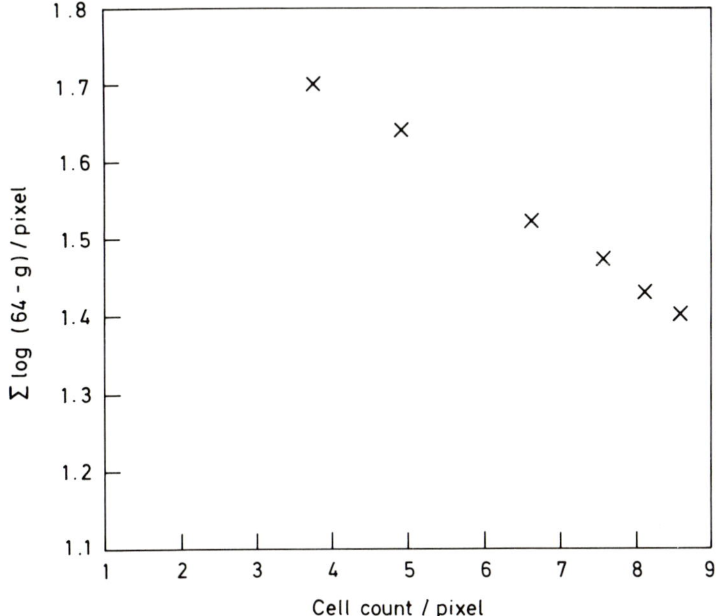

Figure 5.11   Sum of logarithms of grey levels forming a colony per pixel against number of cells in the colony per pixel.

The conclusion is that the the sum of the logarithms of the grey levels forming the colony is proportional to the number of cells in the colony.

9. FURTHER APPLICATIONS

In addition to the use of CLIP4 for simple counting and analysis of colonies, a more detailed investigation of measurements relating to the colonies in the dish as a whole was made. It was hoped to combine the total number, the mean area and the mean integrated optical density of the colonies in a dish into a measure of how well that particular culture was growing.

Experiments were set up using samples of the same bone marrow with different doses of stimulants. All product combinations of the parameters were plotted both arithmetically and logarithmically against the dose of the stimulant (either arithmetically or logarithmically). The best

results were obtained by plotting (total number of colonies
x mean colony area x mean colony integrated optical density), or CAD, against the stimulant dose. Sigmoid curves
were obtained such as that shown in Figure 5.12.   Part (a)
is where there is insufficient stimulant to produce rapid
growth. A linear relation between log (CAD) and log (Dose)
exists at (b), before saturation of the stimulant occurs at
(c).   The experiment has been repeated extensively with
different treated bone marrow samples (FU2, FU4, etc.) and
different types of SA stimulant (PCM – human placental conditioned media, WCM – Wehi 3B conditioned media, etc.).
Similar curves appear with excellent correlation to a
straight line on the centre section (b). Figure 5.13 shows
the straight line sections of the sigmoid curves for different bone marrows (FU2, FU4, FU6) with PCM stimulant and
FU2 with KCM (K5637 conditioned media) stimulant. Separate
lines are obtained for the different stimulants.

The digitisation errors on the log (CAD) values are
never greater than 3%. The implications of this result are
very interesting.   It simply means that the CAD or log
(CAD) value, on the straight line part of the dose-response
curve, appears to give a good indication of how well a particular culture is growing relative to another. Therefore
it is simple to test new batches of GMCSA and SA, or to
investigate differences due to treatment with FU.

With this knowledge, some experiments now being carried
out use this relative measure of growth activity. Four
major areas of research are being undertaken:

1. Batch testing of conditioned media.

2. Assay testing of GMCSA.

3. Comparison of repopulating abilities of dif-
    ferent marrows.

4. Response of colony-forming cells to immunisation
    and subsequent challenge with <u>Salmonella
    typhimurium</u>.

These will be briefly described.

9.1 Batch Testing of Conditioned Media

The CAD or log (CAD) number can be used to investigate
comparisons between different conditioned media, using the
same GMCSA but different SA.   Figure 5.14 shows plots of

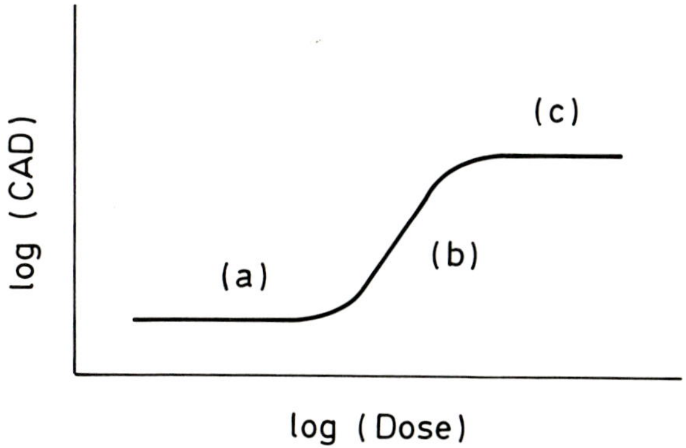

Figure 5.12   Typical log (CAD) against log (Dose) curve.

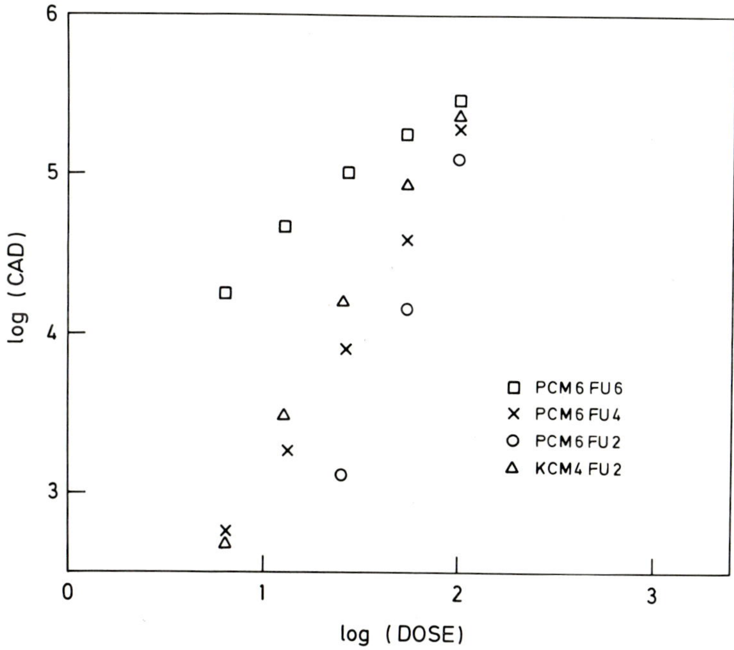

Figure 5.13   Dose response curves for various bone
marrows and stimulants.

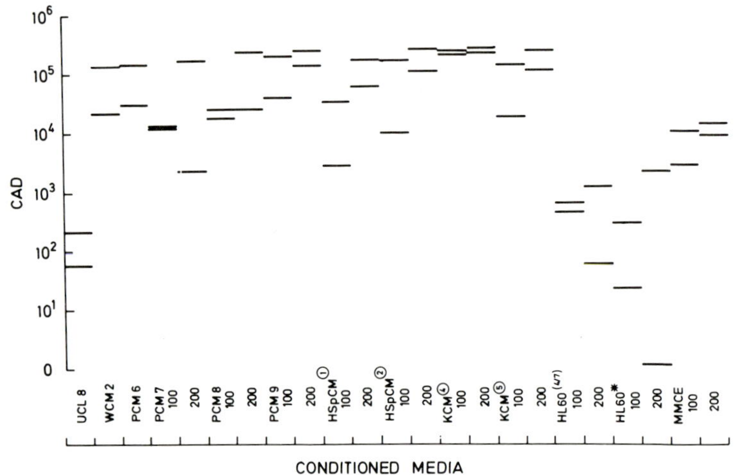

Figure 5.14   CAD for various SA stimulants.

CAD against various concentrations of different SA as well
as a control which has no SA.   Two plots are presented for
each medium;   the higher number is for FU6, the lower for
FU3.   Any medium with a higher CAD value than the control
indicates successful stimulation of the HPPCFC.   The media
can also be compared amongst themselves to find the best
stimulant.

9.2 Assay Testing of GMCSA

When a new batch of GMCSA has been prepared, the stimulant has to be heat treated to denature inhibitors present.
The amount of heat treatment is critical as too much will
denature the stimulant proteins also.   Heat treatment
destroys the hydrogen and peptide bonding of the protein
structures, rendering them useless.   Measuring the value of
log (CAD) for the same conditioned media with a range of
heat treatments applied will lead to an estimate for the
optimum heat treatment time for the rest of the batch.
Figure 5.15 shows two graphs of log (CAD) against heat
treatment for a new batch of GMCSA.   The top graph is with
WCM SA, the bottom is with PCM SA.   The straight lines are
controls of an optimum value of a previously successful

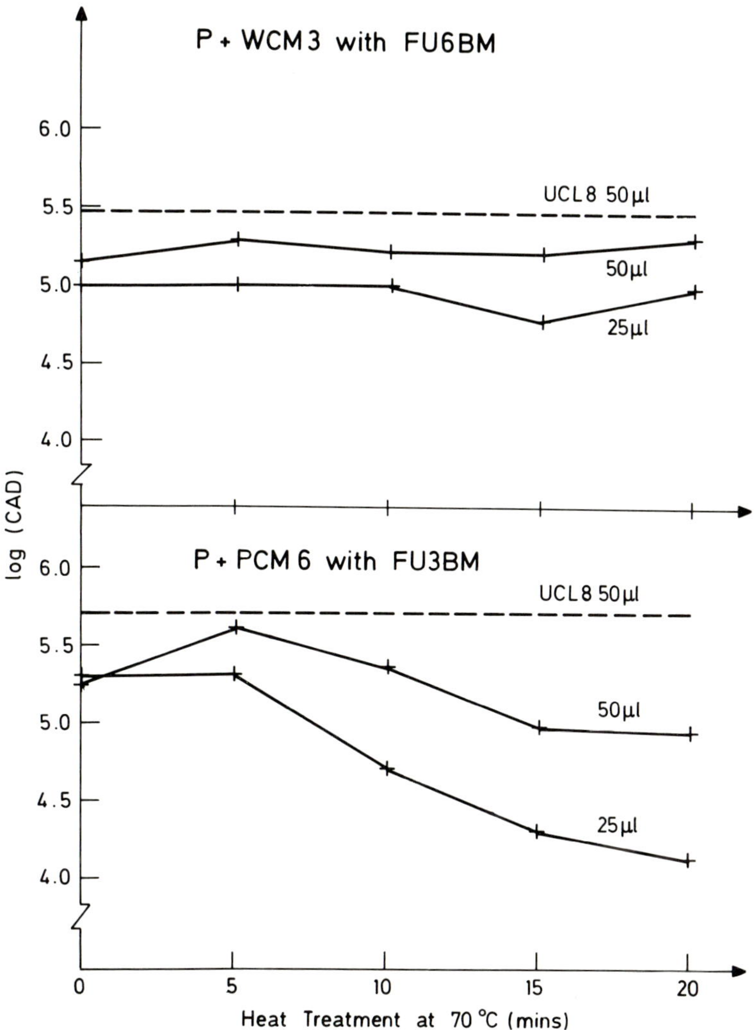

Figure 5.15   *log (CAD) against time of heat treatment for WCM SA and PCM SA.*

batch. In the batches 50 µl and 25 µl of SA have been used.
As can be seen, the optimum heat treatment for all types
appears to be 5 minutes in this example.

9.3 Comparison of Repopulating Abilities of Different Marrows

The ability of CLIP4 to give quickly a value relating to
the growth rate of a culture in response to various stimulants and media leads to a very powerful tool which may have
clinical applications. At present, a transplant patient is
given donor marrow with no knowledge of how well the marrow
will respond. If the machine can give some indication of
which marrow will respond best before a transplant, this
will greatly improve the chance of success.

The latest experiments to see if this ultimate aim can
be achieved involve irradiating mice with a dose sufficient
to kill off the haemopoietic system, and comparing their
dose-response curves with that for normal bone marrow.
Figure 5.16 shows two graphs of log (CAD) against the loga-
rithm of the concentration of two different stimulants for
three sets of bone marrow: normal, 10% effective (i.e. 90%
of total marrow cells killed) and 1% effective. Higher CAD
values are observed for the normal bone marrow, followed by
the 10% effective, and lastly the 1% effective, as expected.
This further confirms the effectiveness of the CAD measurement.

9.4 Response of Colony Forming Cells to Immunisation and Subsequent Challenge with Salmonella Typhimurium

This work has been carried out in collaboration with
Dr B Wilson of St Mary's Hospital, London and Miss J Adam of
the Department of Anatomy and Embryology, University College
London.

Balb/c mice are very susceptible to Salmonella typhimurium: less than a hundred organisms are sufficient to kill
50% of the mice. Immunisation intra-peritoneally with
autologous lipopolysaccharide, extracted with trichloro-
acetic acid, has been shown to protect these animals. The
capacity of precursor cells in bone marrow to form G or Mφ
in vitro was studied in immunised and non-immunised mice
after challenge with S. typhimurium. Femoral marrow was
fractionated at unit gravity and three cellular fractions

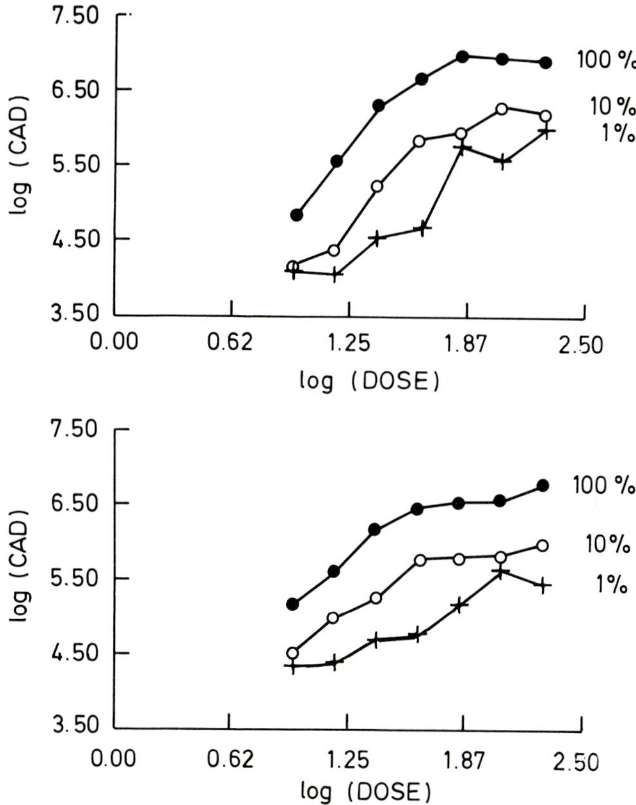

Figure 5.16   Dose response curves using two different stimulants for three types of bone marrow - normal, 10% effective and 1% effective.

assessed in terms of the number and size of G and Mϕ colonies they formed in culture.   The mean colony size in non-immunised mice was found to be smaller than that in immunised mice in any corresponding pair of fractions either 24 or 96 hours after infection.

The hypothesis being tested is that these findings represent the ability of immunised mice to use their developmentally early colony-forming cells (with the capacity to form large colonies) after a shorter lag period or more rapidly after challenge than non-immunised mice.   For further details see [37].

10. CONCLUSIONS

A complete computer system has been developed for the analysis of stained colonies in culture dishes. The system counts the number of colonies in a dish and finds their individual areas and integrated optical densities. The machine is accurate to a sufficient standard and is as fast as required at present. Analysis of one hundred dishes takes approximately thirteen minutes (including human interaction) compared with six hours using the eye and excluding the density measurements. Some experiments can have as many as six hundred dishes and the use of CLIP4 replaces the long and boring process of analysis by eye. More experiments can be set up as the analysis is so much faster and the group at University College London now completely relies on CLIP4 for the analysis of dishes.

Furthermore, a means has been devised which appears to provide a relative quantification as to how well cultures grow. The implications of this have been described. Future work includes the use of the new CLIP4S machine, a scanned array allowing analysis of a 512 x 512 pixel picture, improving resolution so that smaller colonies can be analysed.

CHAPTER SIX

COMPUTER TOMOGRAPHY

The last of the three medically-related projects is in a rather different category. There are several commercially available processors specially designed to carry out the calculations necessary to reconstruct tomographic data. It was a challenge to see whether a general-purpose array such as CLIP4, originally intended for image analysis, could be used effectively to carry out a reconstruction. If it were found to be so, then a further bonus would be that the reconstructed image would then be located in a powerful image analyser and could therefore be immediately, and at no additional cost, subjected to enhancement and analysis.

Keith Clarke worked on this problem, originally in collaboration with the Nuclear Medicine Department of Southampton General Hospital, and the following Chapter is taken from his thesis 'Reconstruction of Nuclear Medicine Images on the CLIP4 Computer'.

The process of tomography may not be familiar to some readers and not all the terms used in this Chapter are explained. In the work described, attention was concentrated on emission tomography. For a clinical application, the subject is injected with a radioactively tagged substance which passes into the blood stream and then flows into various parts of the body (selected substances are used to investigate particular organs or regions of the body).

Consider all the radiation from a horizontal slice through the chosen organ. If a collimator is placed due north, say, of the horizontal slice, then all the gamma rays emerging from the slice in a northerly direction can be collected (after optical frequency conversion) on a strip of photographic emulsion. A small region of the strip therefore integrates the gamma rays from a thin pencil through the slice in the northerly direction and this intensity is referred to as a ray-sum. The linear strip of intensities is called a projection. Projections are then collected at intervals of a few degrees by rotating the collimator-camera assembly about a vertical axis through the organ. The set of projections for all angles is a sinogram.

The process could be reversed by replacing the camera by a projector, so that light shines through the projections. By summing the intensities from projections at all angles, an image of the previously unseen cross-section could be built up. This operation can be emulated on a computer.

In practice, the sinograms are usually recorded electronically and a certain amount of image filtering is required to remove unwanted artefacts, but the principle of reconstruction remains essentially as stated. A more detailed explanation of the various types of tomography now in use can be found in Keith Clarke's thesis and in the extensive literature which has appeared in recent years.

1. INTRODUCTION

The reconstruction methods used in computerised tomography require some basic operators that must work on large amounts of data. Their implementation is the major influence on the speed of processing. The CLIP4 machine's cellular architecture provides very fast processing for suitable operations on whole images. Some parts of the image reconstruction procedures map directly on to the architecture, while others require novel algorithms to ensure good performance. The techniques that have been developed to produce two complete parallel beam reconstruction programs on the CLIP4 system are described here and an analysis of their performance is given.

2. CHOICE OF ALGORITHMS

The major operations required for analytic and iterative reconstruction are summarised in Tables 6.1 and 6.2. The choice between the algorithms will depend on the quality of the results they generate, and whether they can be executed in an acceptable time. The quality of the results reflects how well the assumptions used to derive an algorithm and its discrete implementation fit the practical data. The time taken to produce the results depends on the required operations being efficient on the computer system used. With conventional serial computers it is possible to assess times by considering the total number of adds, multiplies, etc. needed and scaling them by the instruction times. As soon as more than one processor is involved the partitioning of the processing becomes important. CLIP4 is an extreme case where an operation on a whole image of data takes the same amount of time as an operation on a single pixel of data.

On CLIP4, moving individual pieces of data to different positions makes no use of the machine's parallelism. The backprojection and reprojection operations, manipulating sets of ray-sums, are not so arbitrary. The former is a transform from one to two dimensions and the latter from two dimensions to one. The relationship between adjacent points is maintained as they are spread back or accumulated. An important consideration is the number of bits used to represent the data. An upper limit is imposed by the amount of bit-plane memory in the system. Numbers with a large number of bits are time-consuming to process because the bits are treated serially, possibly imposing a limit on the precision used.

Table 6.1   Summary of Fourier-based analytic reconstruction methods

Technique	Operation
Fourier filtered backprojection	FT_1 scale by ρ FT_1^{-1} backproject sum backprojections
Fourier filtered summation	backproject sum backprojections FT_{12} scale by ρ FT_{12}^{-1}
Frequency space interpolation	FT_1 interpolate FT_{12}^{-1}
Convolution filtered backprojection	convolve with $(FT_{12}^{-1}\,\rho)$ backproject sum backprojections
Convolution filtered summation	backproject sum backprojections convolve with $(FT_{12}^{-1}\,|\rho|)$

Consider the first three algorithms in Table 6.1. They rely on the explicit computation of Fourier transforms of the data. Spatial frequency components are selected by convolving the data with sine and cosine waves. The transform of real, general data has both a real and an imaginary part. The amount of computation can be reduced by making use of the fast Fourier transform [38]. Some use of parallelism can be made because all rows or columns can be transformed together; a two-dimensional transform can be implemented as a one-dimensional transform on all rows or columns followed by a one-dimensional transform on all columns or rows. Otto [23] considers the Fourier transform on CLIP4 and points out that the time taken is dominated by the arithmetic rather than by the data-shifting operations. The Fourier coefficients will vary by several orders of magnitude so that floating-point methods using many bits are appropriate. The Fourier transform methods are therefore unattractive on CLIP4.

The two remaining methods in Table 6.1 use convolution operations to perform the filtering. Once a set of numbers representing the inverse Fourier transform of a windowed ramp function have been calculated, they can be used for all sets of data. The windowed ramp function has a real, spatially limited, symmetric transform. The convolution operation is straightforward to implement as a series of shift, scale and accumulate operations. Of the two methods, the convolution filtered backprojection requires less operations and precision; filtering before backprojection and in one dimension is less demanding, in terms of both number of operations and precision, than filtering the two-dimensional image after backprojection. The analytic method of choice is therefore convolution filtered backprojection, which will be referred to from now on as filtered backprojection (FBP).

The differences between the iterative methods of Table 6.2 mainly concern the order in which corrections are obtained and applied. ART and MART obtain a whole reprojection for a particular direction at once and add in or multiply in the whole backprojected correction to all trial image points. ILST and SIRT apply the correction to single trial image points in turn so that little use can be made of parallelism. The multiplication and division required in MART for the correction step is much more demanding than the simple addition required in ART. The iterative method of choice is therefore ART.

Concentrating on only two methods is not as limiting as it might seem at first. The FBP technique is perhaps the most popular and therefore well-researched reconstruction

Table 6.2   Summary of the iterative correction methods

Technique	Operation	
Algebraic reconstruction technique – ART	Take the reprojection Difference with the corresponding projection Backproject and add in the correction to the trial image	For each projection angle
Multiplicative algebraic reconstruction technique – MART	Take the reprojection Divide the corresponding projection by the reprojection Backproject and multiply in the correction to the trial image	For each projection angle
Iterative least squares technique – ILST	Calculate all reprojections Get the contributing projection and reprojection ray-sums Compare them Calculate a new point value which minimises the disagreement	For each point in the trial image
Simultaneous iterative reconstruction technique – SIRT	Calculate the reprojection ray-sums Compare with the corresponding projection ray-sums Calculate a new point value which minimises the disagreement	For each point in the trial image

method. ART converges much quicker than the other methods and many refinements have been proposed to control the quality of the reconstruction. As well as being suited to CLIP4, the methods are probably those that would be chosen first on the grounds of functionality.

3. OPERATOR ELEMENTS FOR CLIP4

3.1 General

Some ways of implementing the operations required for FBP and ART will now be considered. The target resolution is that typical in nuclear medicine. The emphasis will be on using techniques that make the most of CLIP4's architecture. The major requirements for FBP are convolution filtering and backprojection while those for ART are reprojection and backprojection. To quantify the behaviour of the operators the reconstruction of M projections containing N ray-sums each will be considered, producing a cross-section image on an L by L array of processors. It will be assumed that $M \leq L$ and $N \lesssim L/\sqrt{2}$, so that the whole of the projection sinogram can be held in the array at once with some projection width to spare. The origin for the x,y co-ordinates is assumed to be in the centre of the array. Full use will be made of the low-level architecture of CLIP4 to achieve reasonable speed and investigate the potential of the cell design.

If the number of projections is greater than the array size they can be handled in two or more batches. If the projections are obtained sequentially, then each can be processed separately after being loaded into a single row of the array.

3.2 Convolution Filtering

In a general convolution operation, each data point is replaced by a weighted sum of all other data points. At each point, the element i pixels away is weighted by $h(i)$ and accumulated. When the weights are spatially limited to $2R+1$ points and symmetric, the range of i will be $-R \leq i \leq R$ so that each point depends on neighbours from a maximum of $\pm R$ pixels away. Suppose a projection is loaded into the bottom row of the array. The whole projection could be

scaled immediately by h(0) and saved in an accumulator.   A
copy of all data elements could then be shifted to the left,
a parallel operation on CLIP4, weighted by h(1) and added
into the accumulator.   The shifted copy could then be
shifted a further step to the left, weighted by h(2), accumulated and so on.   The total number of operations required
to convolve the sinogram is independent of the number of
elements in the projection, N, depending only on the
'length' of the convolution weights 2R+1.   Furthermore, if
all projections are available at once in a sinogram then all
the projections could be filtered at the same time.   In
total, 2R+1 shift, scale and add operations are required.
Working serially, M*N*(2R+1) scales and adds would be
required ignoring the data address calculations.   The convolution operation is illustrated in Figure 6.1.

To make use of integer arithmetic the convolution
weights must be pre-scaled to integer values, and the final
accumulator result divided by the scaling factor.   Depending on the number of bits used, rounding errors are a problem, particularly as the convolution weights generally have
a positive central value and are negative elsewhere.

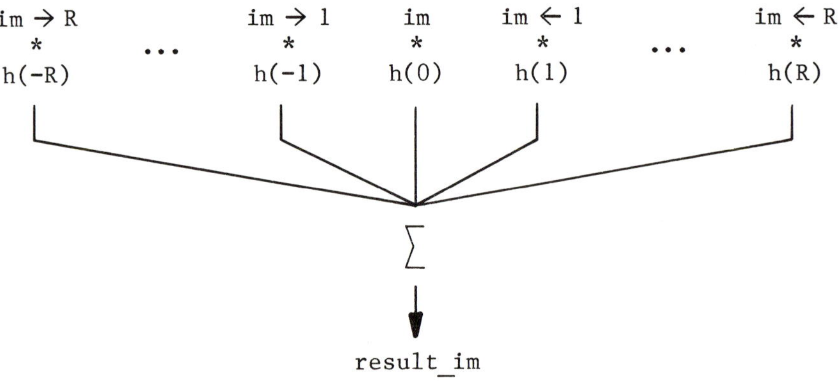

*Figure 6.1   Convolution filtering of rows of data con-
tained in image im.   The operator →i means a shift of the
image i pixels to the right.*

3.3 Backprojection

The backprojection image must be produced from samples
of the projection. A one-dimensional set of numbers must be
smeared across the array at an arbitrary angle and the two-
dimensional image estimated. The first problem is that of
transferring the data from one place to another. Communication is only possible between adjacent processors. If the
smearing were done by local shifts at least L of them would
be needed to copy the line of numbers right across the
reconstruction grid. For example, consider an approximately vertical projection. It must be distributed by
repeated shifts, mainly in a vertical direction but occasionally left or right to follow the required slope. A list
of shift directions would be needed to spread the data
through the correct cells. The high speed of global propagation operations in transferring data from point to point
makes them more attractive for the backprojection task.
Making the propagation flow through the correct cells in a
single operation requires some software mechanism to supple-
ment the cell input gating, which is set to the same direction in every cell. One such mechanism will be discussed
below. Once the data distribution problem has been overcome, the next problem is that of estimating correct values
from the projection samples. For the present, it will be
assumed that the closest projection sample value is a sufficiently good estimate.

Digitised lines

First of all, the cells to which a ray-sum should be
distributed will be considered. A line in a digitised image
is made up of a set of connected pixels. When connected is
interpreted to mean adjacent to any of the eight possible
neighbours, the line is said to be eight-connected. When
connected is interpreted to mean adjacent to the four vertical and horizontal neighbours, the line is said to be four-
connected. The path of a continuous line is approximated by
the set of connected pixels which are closest in some sense.
In the case of straight lines, an eight-connected digital
approximation might be constructed by working out the exact
x co-ordinate for each integer y co-ordinate. The x value
is rounded to the nearest integer and the pixel at integer
x,y set to 1. A set of digital lines is shown in Figure
6.2. The centres of the points on the vertical, horizontal
and diagonal lines lie exactly at the co-ordinates of their
continuous counterparts while the centres of all other lines
fall up to a maximum of half a pixel away.

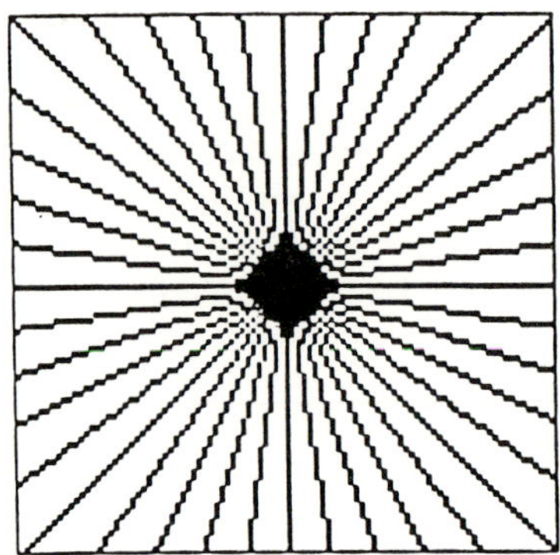

*Figure 6.2   The set of straight digital lines at
9 degree increments passing through the centre of the image.
The points from different lines are ORed together and so
fill in the central region.*

Image (a) in Figure 6.3 is a set of columns of alternate
0s and 1s, dislocated at a particular set of row positions.
Regarding the 1 pixels as eight-connected, they divide up
the image into a very large number of regions of a few
pixels each. If the connectivity is reduced to the neighbours in directions 2, 3, 6 and 7 for both the 1 and 0
pixels then the image can be interpreted as a set of lines
dividing up the image into stripes. The presence of the
lines is clearly seen when just one is selected out; the
points which are 2367-connected to the origin are shown in
image (b). A second orientation, shown in (c), is produced
by changing to 1256-connectivity. The positions of the
column dislocations were chosen here to ensure that straight
lines would be generated.

The 1 pixels in Figure 6.3(a), and a connectivity
specification, define half the possible parallel straight
lines in the image for a particular angle. The other half
can be obtained by inverting the image. Sets of parallel
lines for other orientations can be defined by similar
images and connectivity specifications.

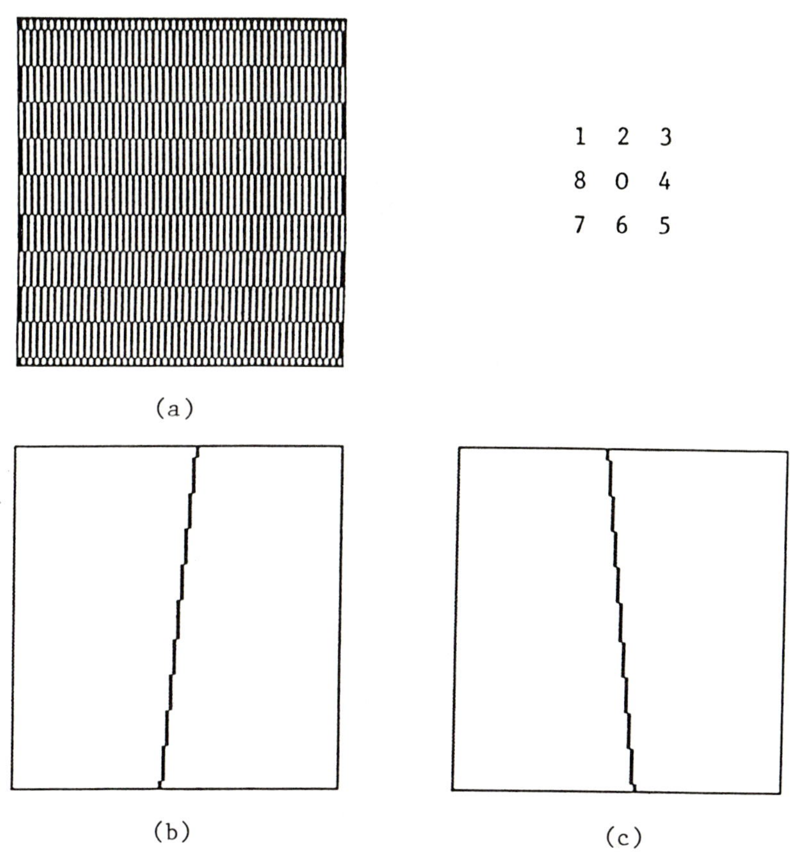

Figure 6.3   Lines extracted from an image by restricting the connectivity.

Labelling on CLIP4

Labelling is a binary operation in which a marker point or points in one bit-plane selects connected regions in another. The mechanism used is a single global propagation operation. The marker points introduce propagation which passes through connected pixels until the boundaries of all connected regions have been reached. For example, when the points in Figure 6.4(a) are used to label the image in 6.3(a) using 2367-connectivity, only the lines in 6.4(b) are retained. The global propagation has been confined to the lines which are separated from each other by a single 0 pixel and reduced connectivity. A second, different set of lines, 6.4(c), is retained when the inverted version of

                    K. A. CLARKE

6.3(a) is labelled.   The OR of the two sets produces the
image of 6.4(d).   In two global propagation operations (the
labellings) and a single pointwise operation (the OR), the
data of Figure 6.4(a) has been spread at an angle right
across the image.

The operation shown in Figure 6.4 is a type of backprojection of the binary marker points.   The same technique
can immediately be used with grey-level data by binary backprojecting the bits in turn.   This **label_lines** technique is
the basis of the fast backprojection operation for CLIP4.

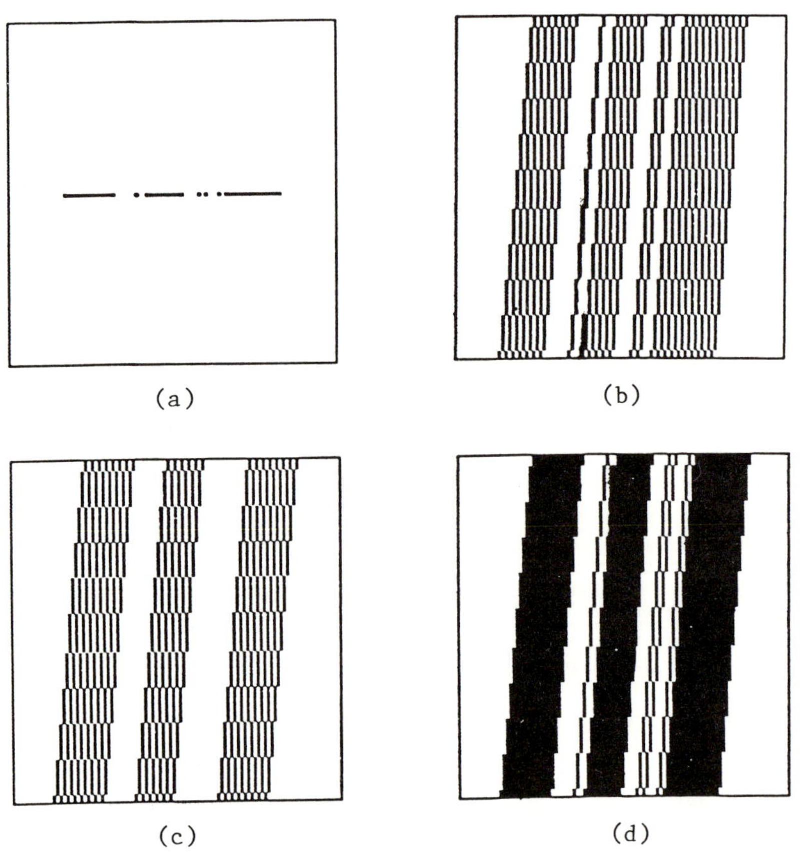

(a) (b)

(c) (d)

*Figure 6.4   The points of (a) label the points shown in
(b) and (c) from Figure 6.3(a) and 6.3(a) inverted respectively when directions 2, 3, 6 and 7 are enabled.   Their OR
is shown in (d).*

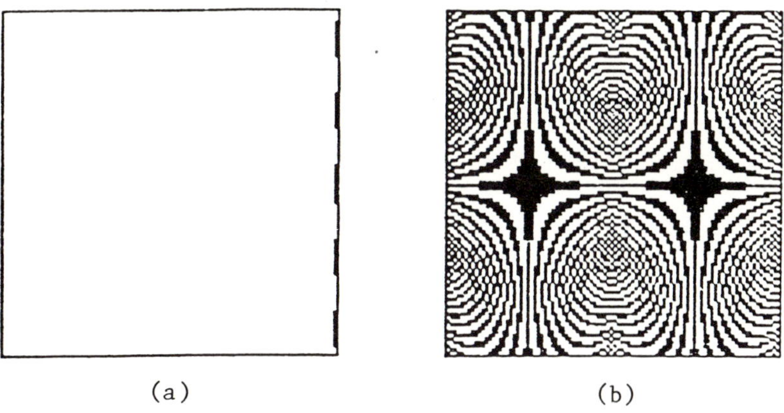

(a)   (b)

Figure 6.5   One row of Figure 6.3(a) and a set of rows of angle data, called the slope_im.

The angle information

The information required to generate Figure 6.3(a) is the set of positions at which the lines should jump one pixel. Working from left to right of 6.3(a), each column is the previous column inverted. The single row of points of Figure 6.5(a) are therefore sufficient to specify the whole image. The label_lines algorithm can reconstitute the whole image by a single horizontal global propagation operation in which the propagation is inverted at each cell. As only one column is required per angle, the information for many angles can be held in one bit-plane. The data in Figure 6.5(b), the **slope_im**, specifies sets of lines at 1.875 degree increments over 180 degrees. When any line is required, it is masked out and the alternating pattern generated.

3.4 Reprojection

Reprojection is the numerical simulation of the projection gathering process. A reprojection is a one-dimensional set of ray-sums, each of which is the sum of a strip of a two-dimensional image. The appropriate contributing points must be selected and added, so the labelling operation is inappropriate. The reprojection operation involves more computation than backprojection because values have to be summed rather than just copied. Working on a digital square grid complicates the process. Consider the

reprojections of a circular region of unit intensity, which
should all be identical, obtained by summing pixels. The
projection vertically downwards will have a maximum value
equal to the diameter of the circle. The maximum of the
reprojection at 45 degrees will be reduced by a factor of
$1/\sqrt{2}$ because there will be that many less 1 pixels. At
some stage compensating weights must be introduced.

Two basic reprojection algorithms were implemented and
both will be described.

A hard way

The first algorithm used mainly local operations. To
collect the data, at least L shifts are once again needed,
and to sum it, L adds would appear necessary. The adds
will become more expensive as the number of bits in the sums
grows. The number of adds needed and the complication of
control flow were reduced by splitting the reprojection into
two operations. The first operation skewed the data on to
the vertical axis by shifting sections of the image left or
right, effectively removing the slope. Masks to select
which parts were shifted were generated from the same angle
data image, the **slope_im,** used for backprojection. The
summing operation then worked by splitting the image in
half, superimposing the two halves by shifting one of them
L/2 pixels, and adding. The half image was then split in
half again, one half shifted L/4, the parts added and so on
until the data had been reduced to a single row of numbers
which were the approximate reprojection. The number of adds
required was reduced from L to O(log L).

An easy way

In a pointwise add, the sum is produced by one of the
Boolean processors and the carry by the other Boolean processor in each cell. The carry output not only goes to the
cell's carry bit, but is passed on to the neighbouring
cells. Bit-column addition propagates the carry from the
sum of two bit-planes containing bit-column numbers across
the array in the direction from the least significant bit to
the most significant bit. With the correct sequence of
operations, detailed in [23], the cell can be configured to
perform a running addition of successive bit-stack numbers
in a particular direction. Furthermore, the propagation can
be limited to lines in the same way as the labelling
described above, so that reprojection summing is possible.
Using the data image (Figure 6.3(a)), with directions 2 and
3 enabled, results in the sum along the lines at the bottom
left edge of the array, and intermediate sums right across

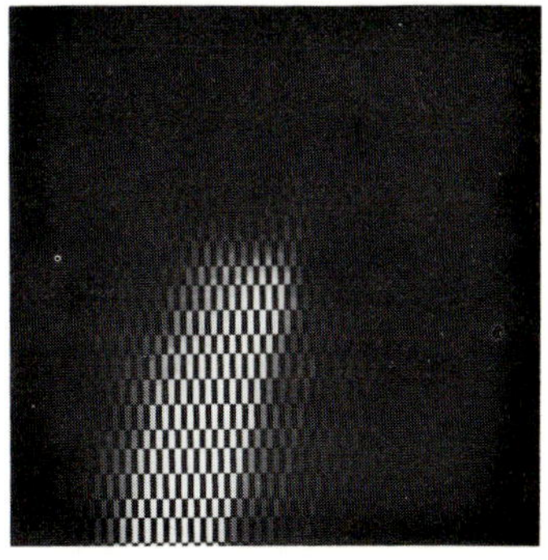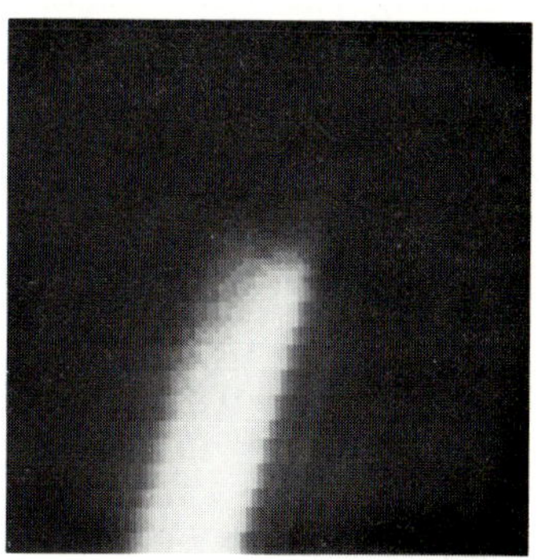

Figure 6.6   One set of running sums along lines and the two sets combined to produce a full reprojection.

the array. Sufficient result bit-planes must be calculated for the biggest possible result image number. Only one global propagation operation and a few pointwise operations are required per result image plane. Once again, half the lines are dealt with at a time, as illustrated in Figure 6.6, and the separate results ORed together.

The benefits of the second reprojection technique are twofold. Firstly, many fewer operations are required because global propagation is used to pass information across the array instead of many local operations. Secondly, very little code is required to perform the reprojection. The many mask generation, shifting and copying operations are replaced by global propagation constrained to flow in the correct directions. The second reprojection method, the **sum_lines** method, is the analogue of the label_lines technique, sharing the same confined propagation but requiring an obscure set of cell functions.

The reprojection process generates a one-dimensional line of large numbers. When stored as a bit-stack most of the stack is empty, so space is wasted; also arithmetic operations are only one-dimensionally parallel. These problems are overcome by making use of bit-column representation. As the lines of sum bits are formed they are shifted into a single bit-plane as bit-column numbers of up to L bits.

3.5 Moving the Data Around

In the backprojection labelling operation the bits of a projection were assumed to be positioned in the middle row of the array. For projections orientated approximately horizontally, the bits would have to be positioned in the middle column. The correct projection from the sinogram must be selected with a masking line and then moved and possibly flipped through 90 degrees. The technique adopted for this positioning is illustrated in Figure 6.7 for a bit-plane of a projection which must be turned through 90 degrees. The required projection is selected by an horizontal cursor line. The masked projection is projected downwards and then the bottom line of the array is masked out. The new image is then projected at 45 degrees and the left-hand edge masked out. An horizontal projection is generated, the centre column of which is selected by another mask image. The masking lines for the centre row and columns, generated by forming a line at the edge of the array and shifting it an appropriate number of times, are calculated once and used for all projections.

To accommodate any projection angle, 180 and 270 degree flips are needed. A 270 degree flip can be accomplished by first propagating diagonally to the right-hand edge of the array and then horizontally, and a 180 degree flip can be done with two diagonal propagations followed by a downward

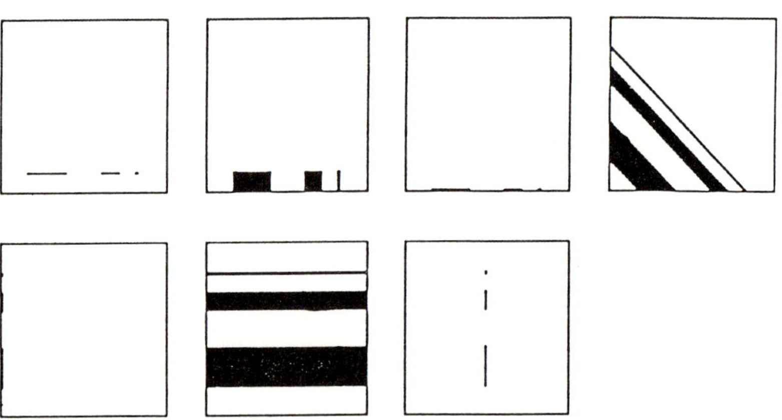

Figure 6.7   The sequence of images produced by moving any horizontal row of data to the centre column.

propagation. The combination of global propagations and cursor lines for moving the data around will be called the **move_row** algorithm. The algorithm is used to position any projection in the centre row or column at an appropriate orientation. The correct line of the **slope_im** of Figure 6.5(b) must also be positioned, this time at the edge of the array, and orientated correctly.

4. GEOMETRIC CORRECTION

4.1 Requirement

The backprojection and reprojection operations effectively work by skewing the data. The width of a projection at angle θ is compressed in backprojection to

$$s(\theta) = \max(|\cos \theta|, |\sin \theta|)$$

of its correct width. The width of the reprojection is expanded by $1/\max(|\cos \theta|, |\sin \theta|)$. At some stage geometric compensation must be applied so that correct widths are maintained. It seems sensible to apply the compensation before backprojection, while the data is still in sinogram form, so that all the projections can be treated at the same time. As each projection may correspond to a different angle, the amount of correction will generally be different in each line of the sinogram.

For ART-type algorithms either the reprojections have to be shrunk or the projections stretched before the two are compared. It is computationally advantageous to do the latter, because the projections have to be stretched only once at the start of the reconstruction, while many individual reprojections are generated during the reconstruction. Only the projection stretching will be considered.

4.2 A Fixed Magnification Factor Algorithm

A paper by Weiman [39] suggested an algorithm intended for special hardware which performed geometric operations by manipulating the rows and columns of images. General geometric operations were decomposed into x and y magnifications and shears parallel to the x and y axes. The algorithm, adapted for CLIP4, is described in detail in [28].

The magnification technique was initially used to provide the geometric correction of the projections. The basic technique provides uniform magnification over all rows of the image because interpolation factors are obtained from a uniform ramp image. To provide row-dependent correction, two magnification factors, 33/32 and 9/8, applied to masked rows were used. The number of applications was calculated for each row by finding the i and j for which $(9/8)^i*(33/32)^j$ was closest to the desired factor, with $i \leq 3$ and $j \leq 5$ to limit the maximum number of applications. The algorithm to correct image **im** was then:

1. Copy **im** to **im_copy**

2. If i > 0 for any row, magnify **im_copy** by 9/8

3. Replace rows of **im** by **im_copy** where i > 0

4. Decrement all is

5. Copy **im** to **im_copy**

6. If j > 0 for any row, magnify **im_copy** by 33/32

7. Replace rows of **im** by **im_copy** where j > 0

8. Decrement all js

9. Go to 2 while any i or j > 0

The procedure required up to eight magnification operations and many masking operations to produce an approximate result. The repeated interpolation steps blurred the data more than necessary.

4.3 A Flexible Magnification Algorithm

A magnification algorithm which could precisely apply different factors to the rows of an image was required. Such an algorithm that can apply the correction in a minimum number of operations uses a control image, the **int_im**, to specify how the sinogram pixels are remapped, and a single multiply operation to perform linear interpolation. The effect of the algorithm on an unfiltered sinogram is shown in Figure 6.8. The masked fixed factor technique was replaced by this technique, which will be referred to as the **stretch** algorithm. The stretch algorithm may be directly used for fully interpolating parallel beam backprojection and for fan beam backprojection (a technique in which the component rays in a projection converge at a point).

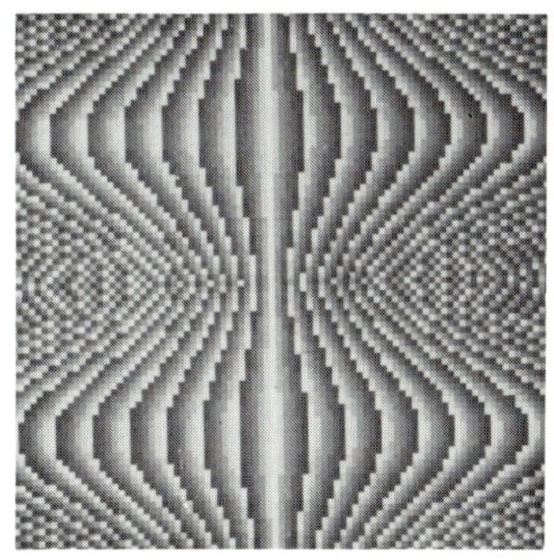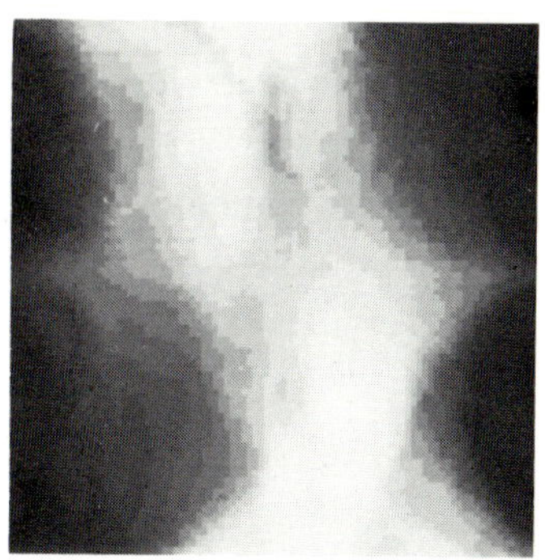

*Figure 6.8   The control image, int_im, and its stretch-
ing effect on a sinogram.*

4.4 Generation of Geometric Data

The backprojection scheme first stretches out the sinogram and then skews the projections back across the array.
Two separate sets of data are required to control the operations.   One set is the binary **slope_im** (see Figure 6.5(b)),
containing the reduced form of the backprojection lines, and
the other is the grey-level **int_im** containing interpolation
factors for the geometric correction. The combined effect
of the stretching and skewing of a projection should produce
an approximation to the backprojection operation. The
geometric data must be precise enough to ensure that the
average slope and position of a backprojected ray-sum are
close to the exact values.

The two data images required are, with $s(\theta)$ as defined
earlier,

$$\textbf{int\_im} = \text{fractional part of } x\left[1-s\left[\theta(y)\right]\right]$$

$$\textbf{slope\_im} = \text{unit plane of } \left[x \tan\left[\theta(y)\right] + \frac{1}{2}\right]$$

where $\theta(y)$ is the angle corresponding to the projection at
row y. Only transition points are required in the **slope_im**
for each θ.   These occur each time the unit bit of x tan θ
changes from 0 to 1 or 1 to 0. Both these images can be
pre-calculated for any set of projection angles, so, for

routine use, a standard set is generated and read in at
run-time. For greater flexibility they can be calculated
at run-time.

5. COMPLETE RECONSTRUCTION PROGRAMS

5.1 General

The techniques described, plus some control flow information, combine to make complete reconstruction programs.
The input data for the programs are sinogram images whose
projections are assumed to cover 180 degrees. The number
of projections is attached in an auxiliary parameter field
of the image header. For convenience in experimental work,
a convention for the direction and angle of the first projection was fixed. Facilities for attenuation correction of
nuclear medicine data are provided by allowing the sinogram
to be scaled by a precalculated weighting image.

Calculations are done on CLIP4 using fixed-point arithmetic. If two n-bit images are added the result will be up
to n+1 bits long; if they are multiplied the result will be
up to 2n bits long. After a few arithmetic operations all
the available bit-plane memory is used up and data starts to
swap on to disc, slowing the program by about two orders of
magnitude. Programs have to scale data down between calculations and the associated truncation errors have to be
accepted. Division is a relatively expensive operation, so
the scaling in the reconstruction programs was restricted to
factors of 2^n only. The division is then simply achieved by
discarding the n least significant bit-planes. Rounding can
be included by adding in the most significant of the dis-
carded planes. The scaling factors used can be accumulated
in the controlling program and correct overall scaling
applied as the last operation by one fully general division.

5.2 Filtered Backprojection

The reconstruction steps for filtered backprojection are
convolution filtering followed by backprojection and summation of all the projections. The image manipulations
involved are summarised in Figure 6.9. The convolution
weights, scaled to integers, are user-selectable from a wide
range of characteristics. All the projections are filtered

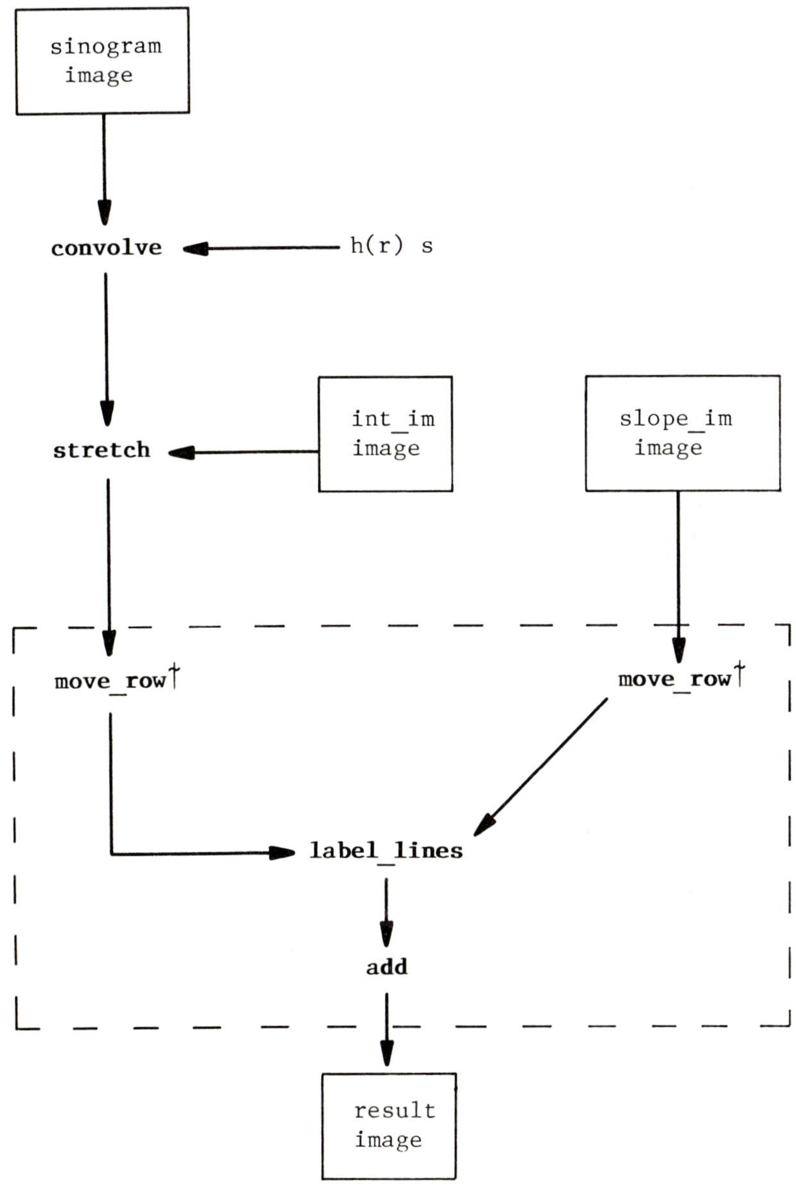

Figure 6.9   The image manipulations in the FBP program. The operations in the dotted box are repeated for each angle.   †A single horizontal cursor line selects the correct sinogram and slope_im rows.

at the same time while they are still in sinogram form. The data images for the number of projections are read in and the filtered sinogram geometrically corrected using the stretch technique. In the auxiliary parameter field of the **slope_im** are three numbers which specify the projections at which the combinations of move_row flips change. A cursor line controls which of the projection and corresponding **slope_im** lines is selected and positioned by the move_row algorithm. The label_lines algorithm reconstitutes the angle data and spreads the projection back. The backprojection is then accumulated into the result image. The cursor line is moved to the next projection row, the next projection backprojected and summed and so on.

5.3 ART

A basic ART program updates a trial image by cycling through all the projections and applying corrections. The image manipulations are summarised in Figure 6.10. Much more one-dimensional processing is done than in FBP because many comparisons between projections and reprojections have to be made. The data images for the number of projections, identical to those for FBP, are read in, and the sinogram stretched so that the effectively expanded reprojections can be compared directly. The sinogram is then weighted in accordance with section 3.2. A trial image, generally a uniform image, is generated. A cursor line selects out a projection and the reprojection corresponding to that angle is produced. The projection is conveniently copied to the edge of the array where the reprojection lies by the move_row and label_lines algorithms. The two are subtracted, and the difference backprojected and summed into the trial image. The cursor line is then moved on to another projection and the process repeated. When all projections have been cycled through once, the process is repeated.

The particular ART algorithm implemented is known as ART3 and is detailed in [40]. It constrains the result to be positive and includes data-dependent relaxation and damping. All the projection data is cycled through three times, with a smoothing operation after each cycle. The trial image at that stage is taken to be the result image. The 'tricks' are parameterised so that the algorithm can be tuned for a particular type of data.

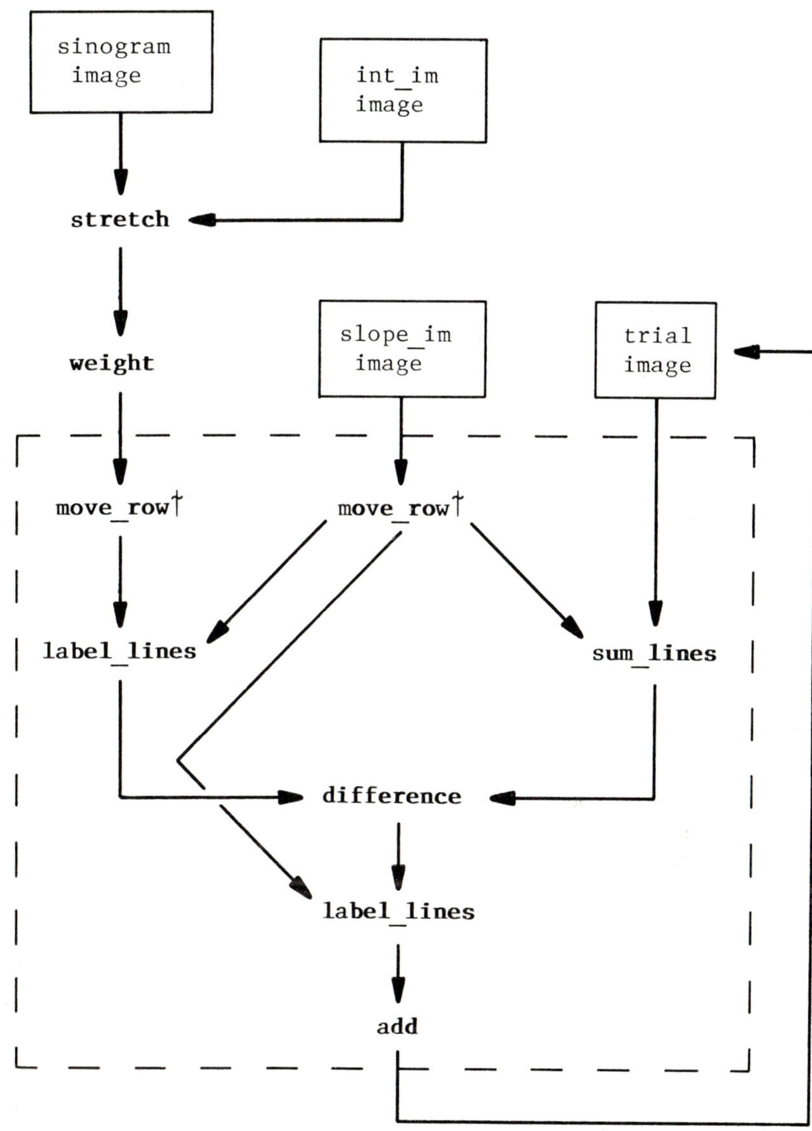

Figure 6.10   The image manipulations in the basic ART program. The operations in the dotted box are repeated for each angle.   †A single horizontal cursor line selects the correct sinogram and slope_im rows.

6. PERFORMANCE

6.1 Speed

A CLIP4 program is a series of array operations controlled by a serial program. The rate at which instructions can be executed is limited by the speed of the array operations and the speed at which the serial program can load instructions into the array. The implementation of the CLIP4 cell logic determines the speed of each type of array operation. The chips used in the prototype machine on which this work was done have a poorer performance than the newer CLIP4D chips which will be used in future machines. Overheads in the serial program controller are also likely to be reduced in future machines.

To broaden the applicability of the speed assessment, three estimates are presented. The first estimate provides a ceiling figure. It was obtained using the high-resolution timer built into the prototype machine and includes some serial program overhead. The second and third estimates are derived from counts of the type and number of array operations executed by the running programs. The figures presented use the instruction times for the prototype chips and for CLIP4D chips. The timing results are summarised in Figure 6.11.

The relative speed of local and globally propagating operations significantly affects the speed improvement with the CLIP4D chips. Global propagation operations were designed into the implementation assuming that global propagation would be much quicker than repeated local propagation, as it is with the CLIP4D chip. While this device runs basically 2.5 times faster than the original CLIP4 chip, the rather greater increase in the speed of global operations results in an overall speed up by a factor of more than five.

The timings do not include the time taken to read in and write out the data. Data access speed on disc-based file systems is limited by head seek-time and data transfer rates; a CLIP4 image might take about 50 ms to read or write. The plots for the ART program, which has to do more calculations per projection than the FBP program, assume one pass through the data whereas typically three passes are needed for adequate reconstruction. ART is therefore more than ten times slower than FBP on CLIP4.

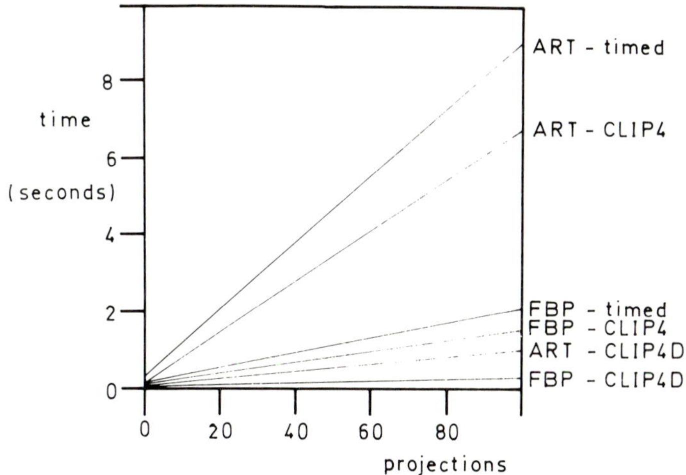

Figure 6.11   Reconstruction times against number of projections for a 6-bit sinogram.

6.2 Qualitative Results

The reconstructions presented here introduce some typical characteristics of the reconstruction methods. The sinogram used for Figure 6.12 was from a simple model consisting of a bright centre circle surrounded by a less bright ring. This data is not typical of nuclear medicine but does make features of the reconstruction easy to discern. When a small number of views are used, as in the left-hand column, the cross-section is poorly recovered. There are insufficient backprojections to approximate the integration over all angles implicit in the FBP algorithm. The backprojection rays are coarsely spaced in angle producing radial streaks. In order to improve the reconstructed image a variety of filter windows can be applied. The responses of three employed in this section, the **rl** [41], the **ga1.0** [42] and the **ga2.2** are displayed in Figure 6.13. The smoothing effect of the ga2.2 filter cancels out the ripples to some extent in the same way as a smoothing operation applied to the completed reconstruction would. With this particular set of data the ART algorithm copes

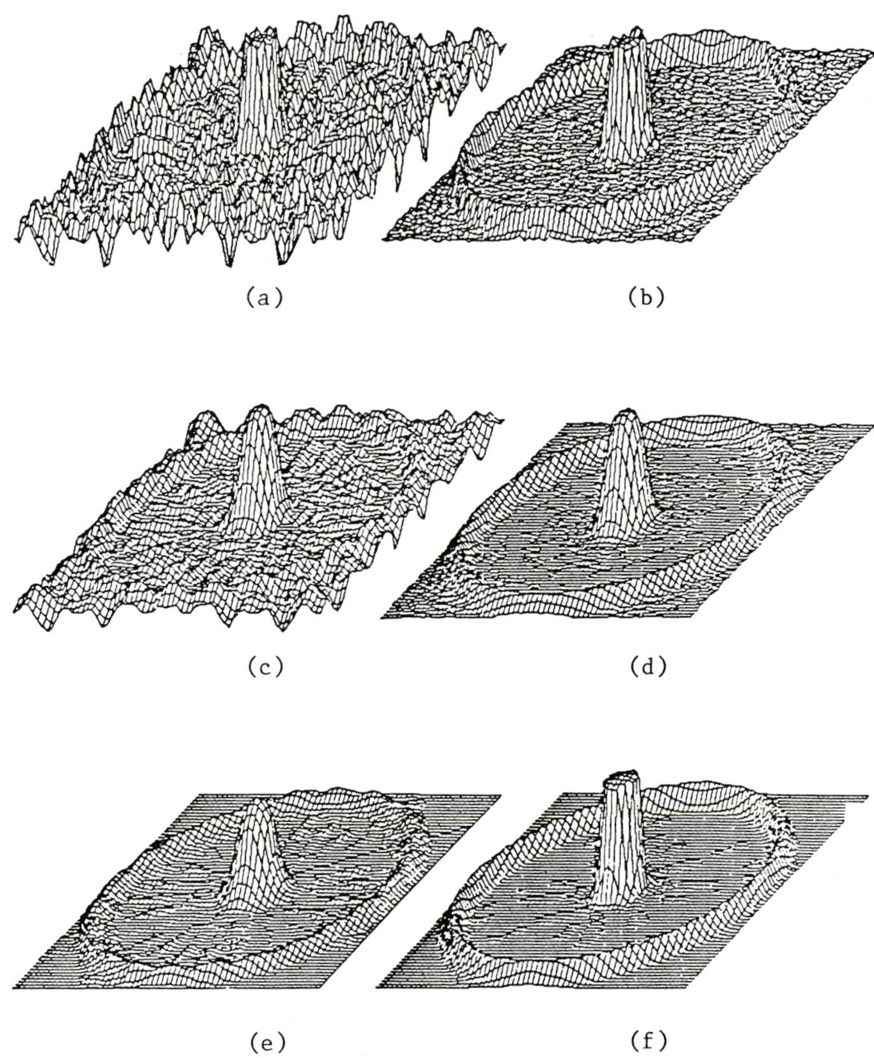

Figure 6.12   Reconstructions of a pillar of diameter eight pixels surrounded by a ring of width 2 pixels of a quarter the intensity.   (a) 12 projections with r1 filter. (b) 96 projections with r1 filter, (c) 12 projections with ga2.2 filter, (d) 96 projections with ga2.2 filter, (e) 12 projections using ART, (f) 96 projections using ART.

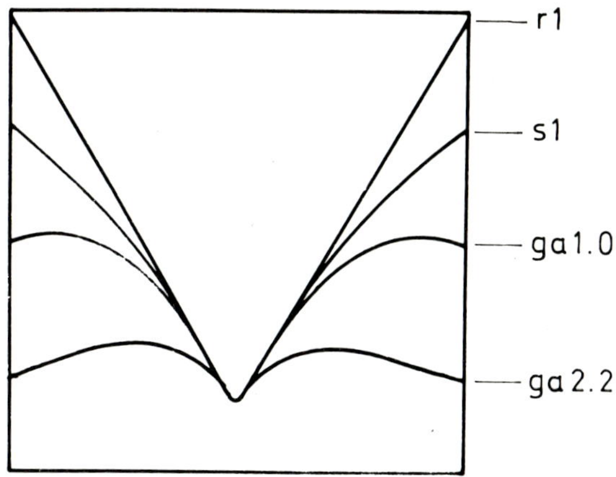

Figure 6.13   A variety of gaussian filter responses used to smooth reconstructed data.

with the limited views the best. The erroneous streaks are inconsistent with the sinogram, and several iterations through the data damps them out.

The reconstructions in the right-hand column of Figure 6.12 were made using a large number of projections. The FBP reconstruction using the rl filter shows some overshoot at the central pillar and noise everywhere. The high-frequency components of small errors in the sinogram are strongly emphasised by the filter. The ga2.2 filter removes the overshoot and noise at the expense of sharpness; a filter with a characteristic between that of the rl and ga2.2 filters would provide a better compromise. In this test, with simple shapes, the ART algorithm produces the best reconstruction, preserving the plateau of the central pillar well and controlling noise well.

6.3 Quantitative Results

The previous section has illustrated that qualitatively reasonable results are possible on CLIP4. In this section some quantitative information is presented in the form of image difference measures. A representative subset of the possible algorithm variations are tested. A reference level is introduced by including the results from a conventional serial implementation.

The measures used for the comparison are taken from [43]. The first represents the sum of differences between the correct cross-section f, and the reconstructed cross-section $\hat{f}$, normalised by the average value:

$$R = \frac{\sum\limits_{\text{all points}} \left| \hat{f} - f \right|}{\sum\limits_{\text{all points}} f}$$

The second measures the sum of the differences squared normalised by the sum of the variances:

$$\delta = \frac{\sum\limits_{\text{all points}} \left[\hat{f} - f \right]^2}{\sum\limits_{\text{all points}} \left[\hat{f} - \overline{f} \right]^2}$$

where $\overline{f}$ is the mean value. The measures generally agree in assessing algorithms, though δ is more sensitive to large errors at a few points than R.

A test sinogram was generated by calculation from a digitised photograph. Table 6.3 summarises the results of reconstructing the test sinogram and the test sinogram with 10% multiplicative noise by the following six different methods:

PDP-11 sl
   A serial implementation of FBP using the **sl** filter [44]
   and linear interpolation in the backprojection step
   coded in 32-bit floating-point numbers. On the PDP-11
   with floating-point hardware each reconstruction took
   about eight minutes of processor time.

PDP/CLIP4 sl
> The sl filtering section of the PDP-11 program passing on the filtered sinogram to the CLIP4 FBP program for backprojection and summation.

CLIP4 ga2.2
> The CLIP4 FBP program using the ga2.2 filter.

CLIP4 ga1.0
> The CLIP4 FBP program using the ga1.0 filter.

ART (smoothed)
> The CLIP4 ART program using selective smoothing between iterations.

ART (unsmoothed)
> The CLIP4 ART program without selective smoothing.

Table 6.3   Global measures assessing the difference between reconstructions and the original data

Method	no noise		10% noise	
	R	δ	R	δ
PDP-11 sl	0.053	0.0018	0.259	0.0346
PDP/CLIP4 sl	0.056	0.0018	0.282	0.0397
CLIP4 ga2.2	0.110	0.0068	0.149	0.0119
CLIP4 ga1.0	0.116	0.0065	0.245	0.0310
ART (smoothed)	0.150	0.0109	0.182	0.0168
ART (unsmoothed)	0.151	0.0105	0.274	0.0432

Table 6.3 illustrates some important features of the algorithms that have been developed, although it does not cover a particularly large set of combinations of data or algorithm parameters. There is a large literature on how filtered backprojection and algebraic reconstruction algorithms perform under different conditions which is directly relevant to the algorithms on CLIP4.

The table is ordered in terms of the R measure for the noiseless case. The PDP-11 reconstruction using the sl filter achieves the best value of R, closely followed by the reconstruction filtered on the PDP-11 and backprojected on CLIP4. The reconstructions done completely on CLIP4 have about twice as much error with this input data. These results suggest that the filtering precision employed on CLIP4 is inadequate; a filtered sinogram backprojected on

CLIP4 is reconstructed about as well as when backprojected
in the conventional serial fashion. The difference in
smoothing introduced by the ga1.0 filter, which has a similar response to the s1 filter and should give good results
with the noiseless data, and the ga2.2 filter appears to
have been swamped by the filtering error. The ART algorithm
performs poorly with this input data.

With 10% noise added, the ranking is significantly
altered. With the data so unreliable, the algorithm's performance under ideal conditions is less important than its
ability to be modified to allow for noise. The CLIP4 ga2.2
reconstruction has the best figures in this case, though the
PDP algorithm could, of course, be modified to use a filter
of similar response. The ART algorithm performs relatively
better here providing that smoothing is applied between
iterations.

7. SUMMARY

The analytic technique most suited to CLIP4 is FBP, and
the iterative technique most suited is ART. An implementation technique for each was developed which attempts to make
the best use of the machine's cellular architecture within
the limitations imposed by the small amount of data storage.
The FBP program takes a sinogram and a user-selected filter,
filters the data, backprojects and accumulates it. The
convolution filtering operation makes full use of the
machine's parallelism when all the projections are available
at the same time. The global propagation feature of CLIP4
speeds data transfer across the array so that the one-
dimensional projections can be quickly spread into two-
dimensional backprojections. The summation of the
backprojections into an accumulator requires only a single
pointwise add. The geometric data for a set of projection
angles can be held in a sufficiently compact form that it
can be read in at run-time.

The ART program takes a sinogram and guesses a trial
image. The trial image is then improved by comparing its
reprojections with the sinogram and backprojecting the
difference. The reprojections are calculated using global
propagation and compared with projections using bit-column
arithmetic. The geometric data required is identical to
that for the FBP program.

The CLIP4 machine is a prototype research machine which has been adapted as new demands have arisen. To separate the processing done in the array from the serial control overheads, an analysis of the array instructions executed at run-time was made. The figures are readily converted to any particular implementation of a CLIP4 processing element. The reconstruction algorithms were optimised in the sense that the processing was arranged to make use of the parallel architecture and fast propagation features of the machine; it is likely that the processing could be completed in significantly fewer instructions by careful low-level coding.

Simulated data were used to provide a basic assessment of reconstruction quality. The CLIP4 reconstructions were compared with ones obtained using a conventional serial program working with 32-bit floating-point numbers. A serious problem when working on relatively noise-free data with the FBP program is its filtering precision. As the filtering step is a relatively small part of the processing time (because all the projections are dealt with at the same time), extra processing would involve little overhead. At the reconstruction resolutions employed, the backprojection and reprojection scheme adopted appears to be acceptable, yielding results which are little different from those obtained by conventional linear interpolation.

CHAPTER SEVEN

MOTION ANALYSIS

One of the impelling reasons for devising new image processor architectures is the need to increase performance to the point where real-time analysis of televised data can be achieved. Since data acquired in this way will often represent moving objects, it is of interest to attempt to find algorithms which will describe the motion recorded in the images. In the past, the relatively slow performance of conventional computers has constrained this type of work to off-line processing and the emphasis has been on accurate description. CLIP4 offers the chance of working on-line and the emphasis therefore shifts from accuracy to speed, algorithms being optimised for this new objective.

This extract from Alan Wood's thesis, 'Parallel Processing Techniques for Image Sequence Analysis', describes studies directed towards this problem and also discusses some novel uses of the CLIP4 array for computations on non-image data. The performance of CLIP4 proved to be insufficient to reach the real-time target but indicated that such methods could be practicable in the near future as faster semiconductor technology becomes available.

CELLULAR LOGIC
IMAGE PROCESSING
ISBN 0 12 223330 1

1. INTRODUCTION

This Chapter presents an application of the tools
described in Chapters 1 and 3 to an actual problem, namely
the analysis of sequences of images. Specifically, the
sequences to be investigated are time-ordered, two-
dimensional images such as might be produced on a cine-film
or by a TV camera.

The purpose is two-fold: to investigate methods for
image sequence analysis, and to use this as a vehicle for
the study and development of parallel processing techniques,
in particular using CLIP4. Thus the work described here
will contain details of the methods used and of the implementation considerations specific to CLIP4.

The main objective of the motion analysis investigation
is to attempt to produce a system which approaches real-time
speeds as far as possible. The goal of the analysis will
be the extraction of objects moving within the scenes
represented by the image sequences. This is therefore not
an object recognition system which implies the existence of
some prior knowledge of the objects to be looked for, rather
it is an object definition system using information derived
from considerations of the properties of motion.

1.1 Problem Domain

The input sequences to be analysed are intended to be as
unrestricted as possible. However, some constraints are
imposed by the available resources such as the imaging systems, or by the size of the CLIP4 array. In the following,
the properties of, and constraints upon, the input images
will be briefly listed, as will an indication of the effects
of using CLIP4.

The images of real-world scenes ·that may be input to
CLIP4 are 96 x 96 arrays of pixels, each pixel being quan-
tised to 64 grey levels. The analysis system will take
these grey-level images as its raw data. Other systems
have used binary images or line drawings as inputs, but here

the decision as to whether to apply this or any other type of data transformation will be left until a later stage of the analysis.

The objects in the input scenes may take any shape, in particular no restraints such as convexity or polygonality will be applied. They may move in arbitrary paths and may occlude each other. Contrary to some other approaches, occlusions in the first frame of a sequence are allowed. Thus there is no guarantee that an object is seen in its entirety at any time.

Each object may be formed from several regions with differing features, such that no segmentation scheme would necessarily be able to merge the regions into the objects using information derived from a single scene. Also, the numbers of regions in each frame of the sequence may vary as a consequence of occlusion and multiple regions per object. There will therefore be no a priori knowledge of the shapes, features or structures of objects that could appear in the scenes.

Neither rotation of objects nor holes within objects are excluded from the input, although the examples to be shown in the following sections will contain neither of these.

Although the paths of objects are unrestricted in principle, the analysis system assumes two-dimensional motion parallel to the plane of the image. Any motion in the third dimension would therefore be interpreted as variation in the observed planar velocities.

The grey level of each pixel of an object is assumed to be constant as the object moves. This implies that for real scenes the lighting and imaging system are uniform or, at worst, slowly varying. The segmentation scheme used will require that the regions of an object be defined by grey-level discontinuities. Thus objects should be separable on the basis of grey level when they occlude.

A set of weak constraints consists of the restrictions which have been applied for practical reasons, but which are not of major significance. Such a constraint is the absence of colour information. If colour inputs were available, it is possible that the segmentation schemes could be improved.

Due to the nature of CLIP4, the pixel resolution of the scene is fixed at 96 x 96. A higher resolution would be likely to improve some aspects of the system and a variable resolution could also be of interest.

No explicit depth information will be available. In fact, the analysis system could result in the derivation of (relative) depth information.

Finally, in order for a multi-region object to be extracted or defined, it must have been visible for at least two consecutive frames. Clearly, no correspondences between objects or their regions would exist otherwise.

1.2 CLIP4-related Features

The subsidiary goal of the investigation, as mentioned above, is the use of CLIP4's parallel processing facilities to permit higher speed processing. Thus properties inherent in parallel processing in general, and CLIP4 in particular, will help to determine the development of the analysis system.

A major factor in the design of parallel programs for CLIP4 has been found to be the need for a re-appraisal of known serial techniques. When faced with a problem, there is often some algorithm or combination of techniques which could be used as a starting point of a solution using a conventional computer. However, these techniques may or may not be applicable to a processor with a radically different structure. Alternatively, some technique previously considered to be too inefficient may be ideally suited to a parallel implementation [45]. The same considerations apply to data reductions or transformations which are often applied to allow an efficient serial algorithm to be used. The influence of a novel architecture can therefore have a positive effect in that it may force an alteration in the conventional modes of thinking.

A more concrete example of the effects of the CLIP4 architecture concerns the limited image memory. In order to use this storage efficiently, different data representations need to be considered, such as a preference for bit-column over bit-stack arithmetic, and arithmetic using integers rather than real numbers.

Lastly, the array size, in addition to defining the pixel resolution, also determines other parameters depending on the use of the array. For instance, the number of regions allowable per frame will be restricted to a maximum of 96 in the experimental system.

2. SEGMENTATION

The main topic of the work reported here is to investigate the way in which regions in time-adjacent frames of an image sequence may be matched.   This has been called the Correspondence Problem.

The segmentation of a grey-level input into distinct areas or regions is required in order to supply data to the correspondence analysis stages.   In other work this has been done in many ways, for example: by using artificially generated objects and scenes [46], by using restricted input domains [47], by analysing on a pixel basis [48,49], and by using an abstracted model of the process [50] as well as using various segmentation schemes on grey-level inputs [51].

The method described here is not an attempt at an optimal segmentation; rather the aim has been to produce an adequate method for the production of data for the subsequent processing, i.e. for an ideal, noise-free input, between ten and sixty regions will be produced which satisfy, as far as possible, the requirements set out below. It is also relatively fast and suited to the CLIP4 architecture.

2.1 Objects

A major goal in the analysis of image sequences is to track objects as they move within a scene.   This may be achieved in two stages:

1. Objects are detected in each frame of the sequence.

2. The detected objects are matched with those found in the subsequent frame, enabling their displacements to be discovered.

The first stage requires a definition of what constitutes an object.   Once defined, objects may be found either by using the definition directly to identify them, or by breaking the scene up into areas which will correspond to parts of objects and combining them according to the object definition.

The first method is likely to be used when specific objects are being sought (e.g. aircraft, cars, missiles). In this case standard pattern recognition methods could be applied to each static scene using prior knowledge of the objects of interest. When a more general input is to be allowed, there will be much less prior knowledge available, making it more difficult to extract objects. Under these circumstances it may be more efficient to use the second method, segmenting the scene according to an abstract notion of what constitutes parts of objects. Then the features of parts may be used to define what is to be looked for in the subsequent frame, followed by an attempt at matching the parts in adjacent frames. Finally, the parts may be formed into objects according to a definition of a generalised object.

Any object definition is recursive and hierarchic: an object may be composed of other (sub-) objects. Consequently, the lowest level of the hierarchy must be specified. In the work reported here, the lowest-level object will be a region as produced by a segmentation process, a region being a connected set of pixels. The object definition will then be as follows:

1.   A region is an object.

2.   A set of adjacent objects moving with equal velocities is an object.

2.2 Segmentation

The task of the segmentation process according to the above definition is to merge image frame pixels into regions which are parts of objects. A region must ideally be part of one object only so that there will be no splitting of regions as objects move away from each other.

In an ideal segmentation of a sequence containing potentially moving objects, a region will be unchanged when it (1) translates, (2) rotates, or (3) occludes other regions, i.e. segmentation and the above operations must commute.

In practice the ideal is unobtainable due both to the discrete nature of the picture elements affecting the segmentation with rotation, and to the fact that occluding regions may have identical segmentation features, causing them to merge. However, non-ideal segmentation can partly be compensated for in the subsequent analysis.

In choosing a segmentation method, knowledge of and/or restrictions on the type of input to be allowed can be used. This has been the case in other work such as that described by Aggarwal and Duda [46] using binary-valued polygons, and Roach and Aggarwal [47] in which a blocks-world input enabled the use of an existing and sophisticated static scene analysis program. However, the more prior knowledge used or restriction placed upon the input data, the less general an analysis system becomes. Thus as little restriction as possible should be applied. Here the objects will be assumed to be moving wholly within the field of view, with brightness differences (contrast) being sufficient to separate occluding regions.

A difficulty arises in attempting to compare segmentation schemes independently from the subsequent processing since their properties will partly dictate the techniques to be applied. For example, a segmentation based on colour may mean a matching using colour would be more successful than matching on texture. A consequence of this is that one segmentation may be successful even if it appears visually worse than another. As a result, the criteria by which a segmentation scheme was chosen were that it should produce regions consistent with the requirements outlined above and should be appropriate to the CLIP4 architecture.

2.3 Mid Slope Boundaries

The basic algorithm used was to take each possible threshold of a scene to give a binary image, produce the edges of this image and OR the result into an accumulator. When all thresholds have been taken, the accumulator is processed to remove isolated pixel noise points and then skeletonised to give single width boundaries.

In practice, the basic algorithm working on images digitised to 64 levels is found to produce a large number of small regions which makes it too sensitive to noise. Two possible remedies are to smooth the input by local averaging, or to apply a more coarse digitisation. In fact, it was found that a combination of the two gave acceptable results; the averaging improving the frame-to-frame consistency with the different digitisation allowing some control over the number of regions produced.

2.4 The Sequences

The full sequences to be used are illustrated in [52]. There are three such sequences which will be referred to by means of the type of objects they contain. Some of their general features will be described here.

The leaf sequence

The objects used in this sequence are taken from TV snap-shots of two leaves. These were processed by a sequence-generating program which uses a description of the paths to be followed by each object and the order in which they are to be stacked if they should occlude.

The leaf sequence has no occlusion and will thus produce identically segmented objects in all frames. The paths taken by the leaves, shown in Figure 7.1, show both slow and fast variations in velocity which will be seen to be significant later. In this, as in all other plots of tracks, the centres of the bounding rectangles of the objects are shown. The bounding rectangle is the shape formed by the x and y projections of the object.

The leaf sequence was used in two versions, segmented with 3-planes and 4-planes digitisation, to investigate the effects of varying the number of regions in a scene.

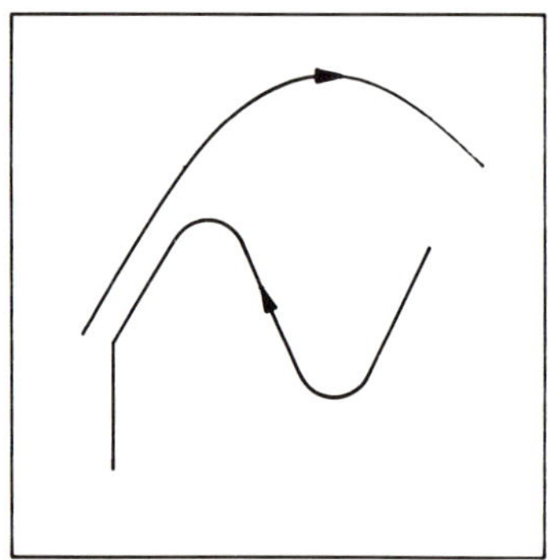

Figure 7.1   Paths of leaf sequence.

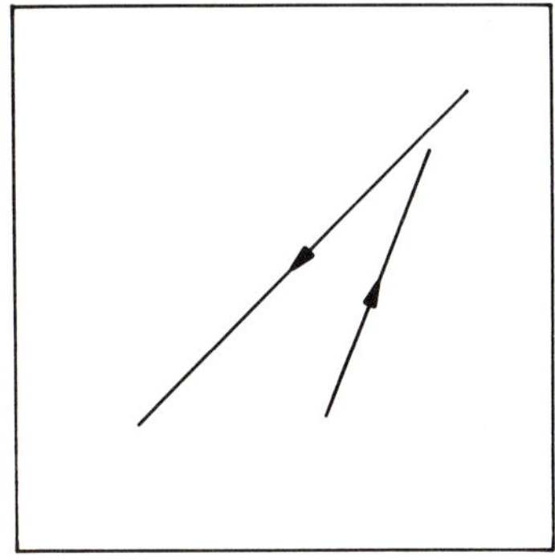

Figure 7.2   Paths of polygon sequence.

The polygon sequence

The polygon-type objects were artificially generated with the grey values of the regions being chosen to be of sufficient contrast to prevent occlusion splitting or merging regions. However, as in the other sequences, the matching process described later is unable to distinguish at least two of the regions. The paths illustrated in Figure 7.2 are simple so as to eliminate effects due to velocity changes. These paths were also adjusted to ensure that the number of regions in each frame of the sequence remains constant, even during occlusion. Thus it is easy to establish the region correspondences that should be produced in the analysis stages.

The objects sequence

This is so named for want of a better description. The objects are real input, using the same TV equipment as the leaves, and are processed to form a sequence whose paths are shown in Figure 7.3.

This sequence also exhibits occlusion, but the use of non-ideal objects means that some alterations in segmentation will be seen when the objects meet. In contrast to the previous sequence, the number of regions per frame alters as occlusion splits and merges regions.

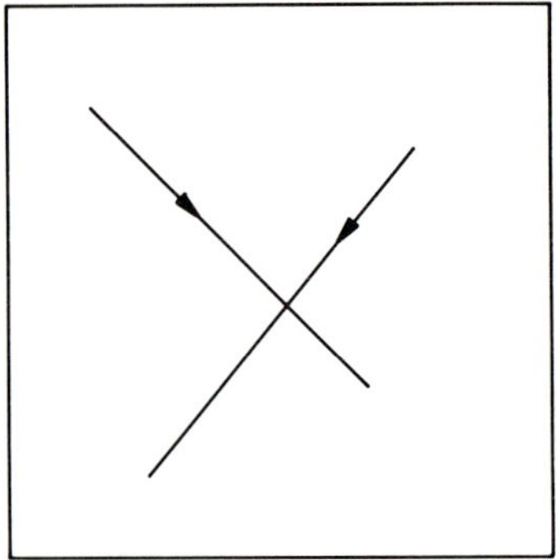

Figure 7.3   Paths of objects sequence.

3. MATCHING

The goal of the inter-frame region matching stage is to
produce a correct one-to-one correspondence between each
region in one frame and its motion-transformed version in
the next, when such a matching exists (an ideal match). In
the case of an ideal segmentation as described above, five
types of region transformation are possible:

1. A region is unchanged except for a possible
    translation or rotation.

2. A region has become partially occluded by some
    other region(s), and may also have translated/
    rotated.

3. A partially occluded region's visible area has
    changed but remains visible.

4. A region has become totally occluded – it has
    disappeared.

5. A previously invisible region has become par-
    tially or totally visible.

The ideal match is therefore an association between regions
in one frame and regions in its successor such that when a
region has undergone transformations (1), (2) or (3) there
is a one-to-one correspondence, but undergoing (4) or (5)

gives rise to no correspondence. The one-to-one matches must of course be correct, matching a region to its transformed image in the next frame.

In practice, an ideal segmentation is not possible even with ideal, noise-free inputs so that, for example, regions may split or merge due to occlusion. In such cases many-to-one or one-to-many matchings would be appropriate. Therefore the list of transformations above is incomplete in practice, although the actions of the ideal matching process for those regions that do undergo such transformations are valid. Also, in practice, an ideal matching process is unlikely to be achieved, leading to at least three possible types of error:

1. Many-to-one or one-to-many matchings may occur which include the correct correspondence.

2. Many-to-one or one-to-many matchings may occur which do not include the correct correspondence.

3. Incorrect one-to-one matchings occur.

The most serious errors are (2) and (3) since they cannot be reduced to the correct matching: error (1) is potentially able to be corrected by deleting the incorrect matches. Thus a goal of a practical matching algorithm must be to eliminate errors (2) and (3), possibly at the expense of increasing the number of type (1) errors. Hence there is a parallel with pattern recognition in which an attempt is made to minimise the number of false classifications whilst leaving any undecided classifications to further processing.

To minimise the possibilities of error, knowledge of the type of input to be expected can be incorporated. As discussed earlier, as the amount of information used increases, the generality of the acceptable inputs decreases. Thus the minimum amount of information should be used consistent with the restrictions on the allowable inputs.

Several different features, and pairs of features were investigated with a view to discriminating the regions in a scene. All were simple pixel-based measures which could easily be extracted using CLIP4.

It was found that the best single feature was mean grey level, especially when partially occluded regions are to be matched. This is to be expected since all the other features are significantly altered by occlusion whereas, if a region has a reasonably uniform grey level, its mean value

will tend to be the same over sub-areas. The segmentation
scheme, being based on a measure of constant grey level,
will therefore tend to produce regions with relatively uniform brightness and is thus well suited to this feature.

The main disadvantage of mean grey level is that it
gives many multiple matches. This is partly because there
is only a limited number of grey values (0-63) which it may
take. Consequently there is a high probability that two
regions will have the same average brightness (being unity
when the number of regions is greater than 64). However,
it can be seen that the set of matches for any region will
contain the correct match except in cases of extreme occlusion, when, say, only one or two pixels of a region are
visible and they have non-average brightness values.

The most promising pair of features was mean grey level
and area. The inclusion of area in the matching process
reduced the number of multiple matches that would be produced using mean grey level alone. However, in strengthen-
ing the discriminatory power, the use of area sometimes
failed to maintain the correct match since it is much
affected by occlusion.

Based on these observations, two possible approaches
could be considered. Firstly, standard matching techniques
could be extended by trying further single features, new
feature combinations or larger sets of features (triplets,
etc.), with a view to improving the mean grey-level results.
Secondly, alternative methods could be investigated to
reduce the number of multiple matches produced using
higher-level information.

Since multiple matches are likely no matter what feature
or set of features is chosen, and it is not clear which
alternative features would be useful and implementable on
CLIP4, the second approach is followed here. Mean grey
level will be used as the single matching feature since it
gives the fewest incorrect matches, is fairly occlusion
insensitive and is easy to calculate.

4. MULTIPLE MATCH RESOLUTION

Given the inevitability of multiple matches, methods for
their resolution are central to the analysis of dynamic
scenes (resolution being the reduction of a set of multiple
matches to the single correct match, if it exists).

Considerations analogous to those discussed earlier apply to this process, however. It is more important that any removal of matches from a set maintains the correct one, than that the match set should be reduced to one, incorrect, correspondence.

4.1 Dynamic Characteristics of Objects

Defining the characteristics of motion to be used in the analysis effectively constrains the type of motion allowable in the scenes to be analysed. So if, for example, low velocity is assumed then this will imply a procedure which, in using this knowledge, will probably be inapplicable to fast moving objects. Such a characteristic could suggest that regions always move a distance less than their length, thus enabling the resolution of multiple matches by deleting from the match set those regions in the following frame that do not overlap their corresponding region in the current frame. However, such a constraint is too harsh in general and it can be seen from the kinds of regions produced by the segmentation that it would be unsuitable here due to the occurrence of narrow regions.

There is therefore a trade-off between the generality of the input domain and the amount of knowledge that may be applied. The following four qualitative characteristics are felt to be a reasonable compromise:

1. Objects tend to move steadily. Paths which have a random motion rarely occur in the macroscopic world. This implies that sudden changes in speed or direction are unlikely and therefore objects tend to move predictably. However, an object must be seen for long enough for its motion to be predictable. Thus, there is an implicit restriction on maximum velocity in order that an object remain in the field of view for sufficient time.

2. Objects tend to have several regions. Generally, an object consists of more than one part. These parts will be of varying brightness, colour, shape, etc. A static segmentation will therefore mark these as separate regions.

3. Spatial relationships between regions of an
    object are constant except for occlusion. This
    assumes that a perfect input is present (noise-
    free, position-independent segmentation, etc.).

4. Spatial relationships between objects change
    slowly. This is a weaker version of the
    previous assumption as applied to objects. It
    implies that either objects are large, or that
    their relative velocities are small.

In all the above, velocity and related terms are compared
with the rate at which frames may be processed. Thus 'slow'
refers to few pixels moved per frame. Similarly, size-
related terms are based on numbers of pixels.

Based on these assumptions, two methods for multiple-
match resolution were devised. The first, predicted
position, is briefly described here; full details appear in
[52]. The second method, adjacency resolution, will be
discussed in detail.

4.2 Predicted Position Resolution

The assumption of steady motion implies that a predic-
tion of where a region will move to, based on its previous
history, will be a good match to its actual new position.
Thus, by calculating the possible positions of a region in
the next frame from its possible velocities implied by its
matches to the previous frame's regions, the multiple match
may be resolved by a comparison of predicted to actual positions.

The method, then, is to note to where in a three-frame
sub-sequence, any first to second frame match would cause
the second frame region to move in the third frame. Then
multiple matches may be resolved by finding which of the
second frame region's predicted positions are closest to its
measured position in the third frame. Of course, the third
frame positions are uncertain when the second frame region
is itself multiply matched.

Algorithm

Consider a three frame sub-sequence: sceneI, sceneII
and sceneIII. In sceneI, there will be several regions
which do not form one-to-one matches with regions in
sceneII. Similarly, there will be several regions in
sceneII which do not form one-to-one matches with regions

in sceneIII.   The assumption made is that the consistent
matched triplet in every case is that in which the displacement between a matched pair of regions in sceneII and
sceneIII is the nearest to that which would have been
predicted by observing the same sceneII region and one of
its possible matches in sceneI. The distance from the
predicted to measured positions of a three frame match
sequence is then noted as a weight for the sceneI to sceneII
match on which it is based.

This procedure is followed for all regions in sceneI
which are part of one-to-many or many-to-one matches to
sceneII. This results in a three-dimensional matrix of
values W, where w_{ijk} corresponds to the distance of the
position of region j (sceneII), predicted from its match
with region i (sceneI), from the measured position of region
k (sceneIII). The minimum values in this matrix then
determine which sceneI/sceneII matches are the best on
grounds of predicted position. By iterating a parallel
global minimum operation, the match sets for sceneI regions
may be resolved.

4.3 Adjacency Resolution

An alternative description of the goal of resolution is
to remove those matches between regions which are less consistent than the alternative matches in a match set. The
previous method defined 'consistency' in terms of the constancy of motion, and was found to fail in situations in
which this assumption did not hold. The method developed
next uses a different implied definition of a consistent
match based on the context of a region, that is the set of
regions to which it is adjacent.

If assumption (2) holds and objects are formed from
several regions, then objects will have inner and boundary
regions. Inner regions are those which are part of a
single object and are adjacent only to other regions of that
object. Boundary regions are additionally adjacent to
regions of other objects. Assumption (3) then implies that
any information derived from the surroundings of an inner
region will remain constant, and assumption (4) allows that
this information for boundary regions will change slowly.
Occlusion affects the situation by changing inner regions
into boundary regions as the boundary regions of one object
move across another.

By looking for the best match between the context of a
region in one frame and the contexts of its corresponding
regions in the next, inconsistent matches may be reduced.

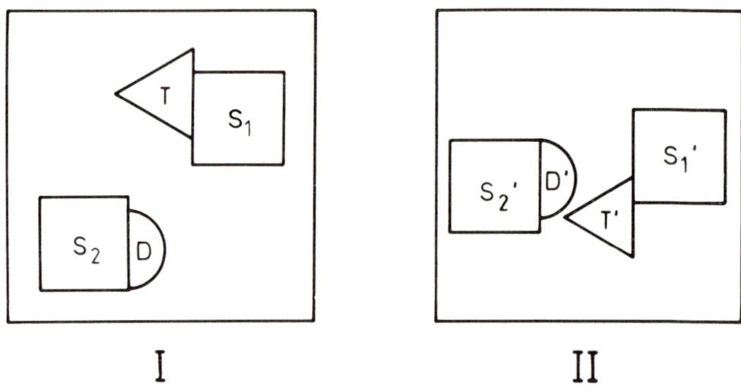

Figure 7.4   An artificial sequence.

Theoretical basis

In the artificial scene sequence shown in Figure 7.4, it
is assumed that matching has been done on the basis of shape
and that consequently the two squares in I are ambiguously
(multiply) matched to the squares in II.   This may be
represented by a matching graph thus:

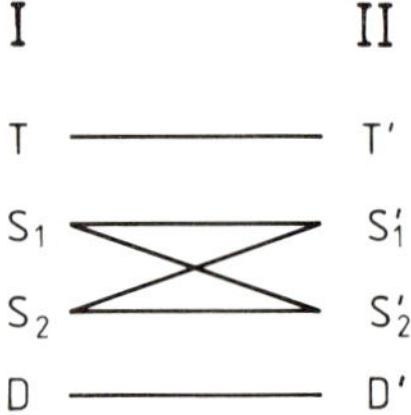

where the arcs represent the relation 'is matched to'.   However, it can reasonably be assumed that S_1 matches S_1' and S_2
matches S_2' due to their different contexts.   Indeed, in the
absence of other information it would be unreasonable
to assume otherwise.   Thus some way of utilising the
contextual information is needed.

The context of a region may also be represented as a graph whose arcs represent 'is adjacent to'. Combining the matching graph and the two adjacency graphs for scenes I and II gives:

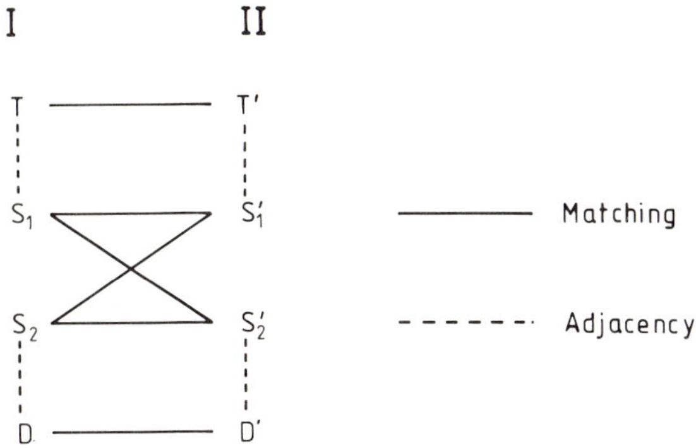

To see how the consistency of a match is calculated using these relations, consider the simplest case:

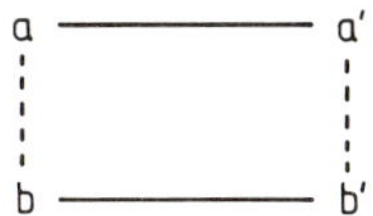

Since a is adjacent to b, b matches b' and b' is adjacent to a' which in turn matches a, it can be seen that the contexts of a and b are the same if the a–a' and b–b' matches are correct. Thus the existence of a loop from a to itself via alternating adjacency-match arcs, ending with a' to a, implies a consistent a–a' match.

In the case of the two frames shown in Figure 7.4, the squares give such loops thus:

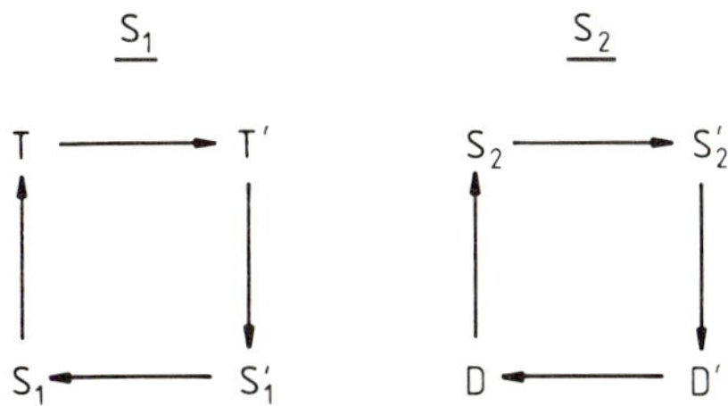

with no loops involving the S_1-S_2' or S_2-S_1' matches.
Therefore, if a weight is given to each match arc, equal to
the number of this type of loop in which it forms the final
arc, the weighted match graph will be:

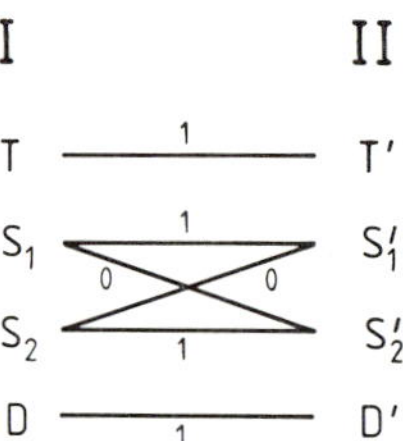

From this the resolved match graph can be formed by
iteratively selecting those nodes with arcs connected to
them which carry the highest global weight. These nodes
and their maximally weighted arcs are added to the resolved
match graph and are removed, with all their arcs, from the
weighted match graph. This continues until the weighted
match graph is empty. The example graph would thus lose
the arcs with zero weights, giving the expected, completely
resolved, result.

If, after this first stage, multiple matches still
exist, the resolution process is repeated until no further
change in the match graph occurs. This is necessary
because removing match arcs from the graph may change the
number of loops of which an arc is a component, thus
enabling ·further resolution. It should be noted that any
exact (one-to-one) match will always be maintained since the
pruning operation cannot remove it even if it has a zero
weight.

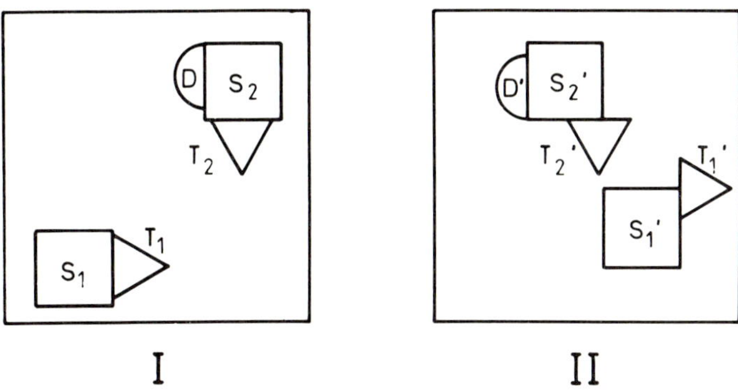

Figure 7.5   Artificial scene pair.

An example which requires two iterations of the resolution process is the frame pair shown in Figure 7.5.   The
weighted match graphs from the two stages and the final
resolution are:

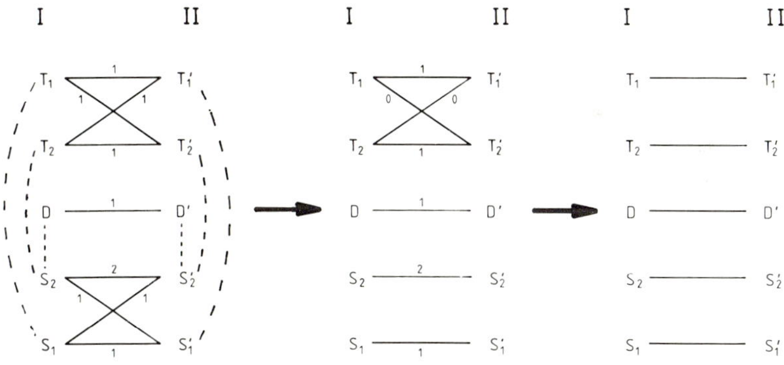

Here it can be seen that due to the unique match of D–D′,
the two squares are resolved in the first iteration.   This
allows the triangles to be resolved in the second.

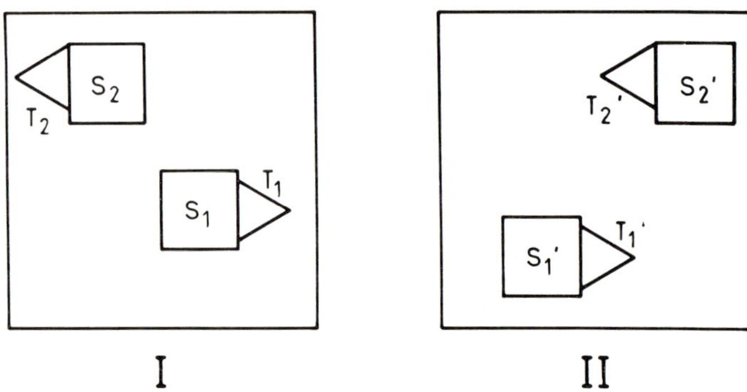

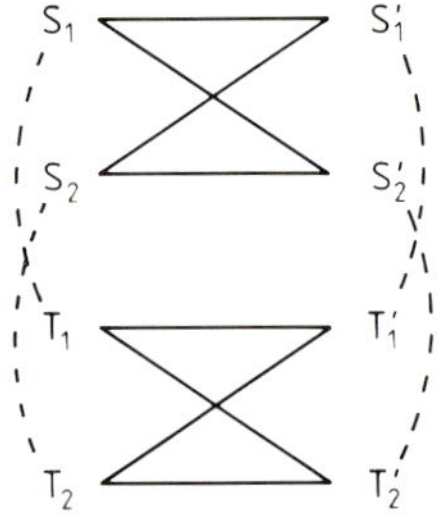

Figure 7.6   Example scene pair.

Two final examples indicate ways in which the method may
fail.   In the first case (Figure 7.6), if a number of
regions in sceneI are multiply matched to the same number of
regions in sceneII, and their contexts are similarly multiply matched, for example:

then no resolution is possible.   This is a 'soft' failure
which future processing, using alternative methods, might be
able to correct.

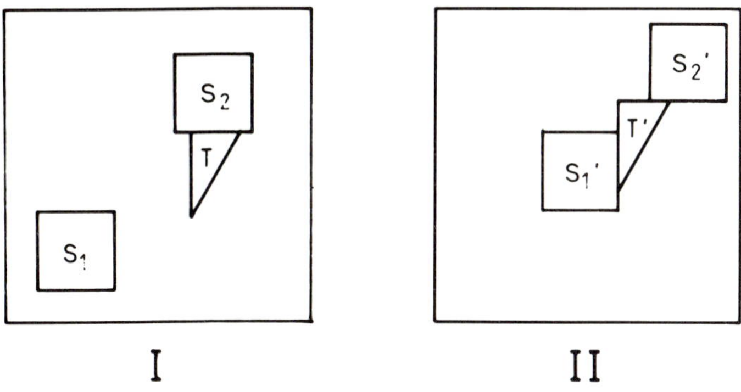

Figure 7.7   Example pair.

A more serious situation is illustrated in Figure 7.7 where resolution causes the following:

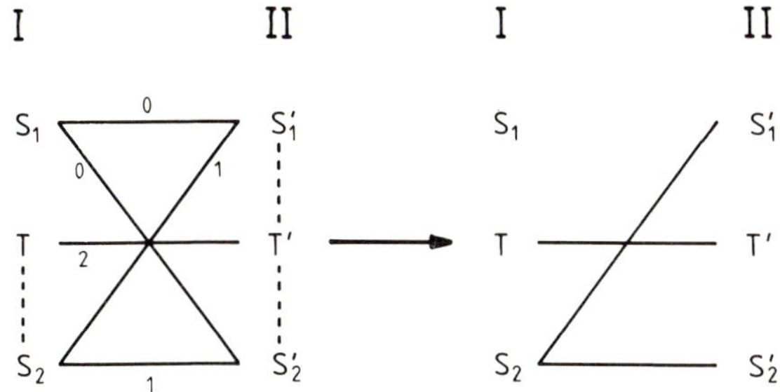

This is a 'hard' error in that the correct S_1–S_1' match has been removed so that no technique which removes matches from multiple match sets can correct this. However, given the constraints and implied definition of match consistency, this is not an unreasonable result as can be seen by examining Figure 7.8 where the small rectangle has disappeared, and the large rectangle (R_2) has split. This example produces the same graphs as above (with S replaced by R) which would appear to be correct under these circumstances.

The main reason for this last failure is that assumption (2) has not held: objects are formed from few regions, object S_1 of Figure 7.7 being a single region. With more regions per object, the information available will be greater and less variable since more regions per object implies that more inner regions will exist, and inner regions will maintain their relationships longer.

                    A. M. WOOD

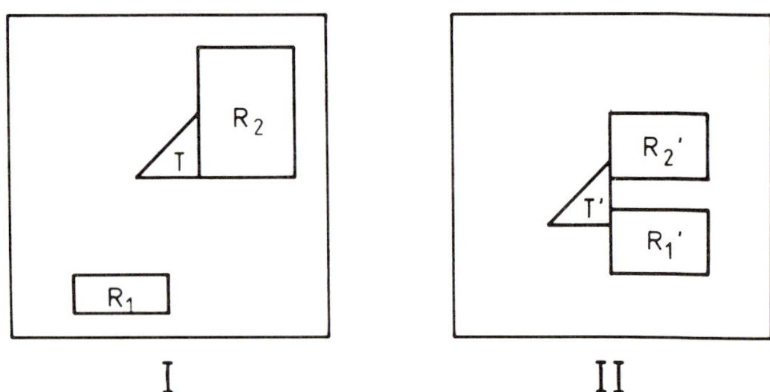

Figure 7.8   Example pair.

This method, then, requires the counting of the number of paths of length 4 consisting of alternating adjacency-match arcs in the combined adjacency-match graph which end on their starting node (4-loops), the assignment of these numbers to the appropriate match arcs, and the pruning of the graph, using these weights as the confidences of each match.

Loop counting

The most obvious method for counting the number of 4-loops for each match arc, namely, to take each arc in turn and follow every adjacency-match path of length 4 incrementing an accumulator when the start node is returned to, is highly serial and time-consuming. Since it is wished to utilise the parallel structure of CLIP4 as far as possible, a much more parallel method must be found.

Consider a portion of a generalised graph, as shown in Figure 7.9, which has been partitioned into three sets of nodes $\{x\}$, $\{y\}$ and $\{z\}$. It can easily be shown [53] that the number of 2-paths from x_i to z_j passing through some node in $\{y\}$ is

$$n_{ij} = \sum_k l_{ik} * m_{kj} \tag{1}$$

where

l_{pq} is the number of arcs from x_p to y_q,
m_{qr} is the number of arcs from y_q to z_r.

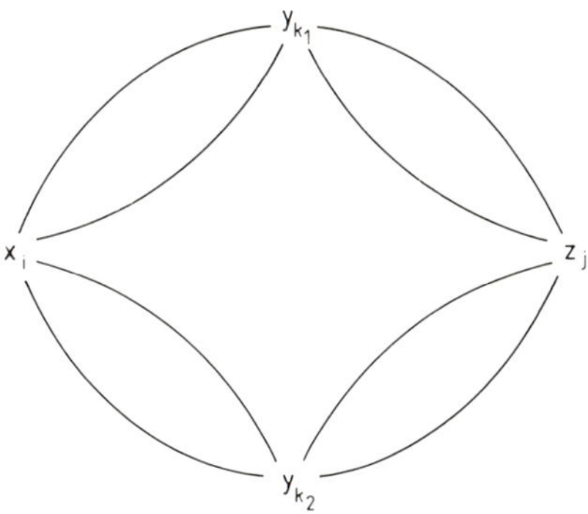

Figure 7.9   Path counting.

A graph may be represented by a matrix whose elements
(ij) are set to the number of arcs from $node_i$ to $node_j$.
Thus the above equation, being the definition of matrix
multiplication, represents the product of the matrices of
two graphs.   The result of such a multiplication will be
the matrix representing the graph with nodes {x} and {z},
having entries equal to the number of 2-paths from a node in
{x} to a node in {z} via some node in {y}.   Therefore, by a
multiplication of n graph-matrices, the matrix containing
the number of n-paths between nodes may be calculated.

Applying this to the 4-loop problem, if adjI, mch and
adjII are the matrices representing the adjacency graph of
sceneI, the I-II match graph and the sceneII adjacency
graph, then the number of 4-paths of alternating adjacency-
match arcs between nodes of sceneI is given by

$$N_4 = adjI * mch * adjII * mch^T$$

mch^T being the transpose of mch.   The number of 4-paths of
a node to itself (4-loops) are then on the major diagonal of
N_4.

Algorithm

The method, as described above, results in the number of
4-paths between nodes without retaining any information as
to their routes.   Since it is required that the number of
4-loops in which specific arcs participate be known, a
modification of the method is necessary.

The sequence of matrix products, adjI*mch*adjII, gives a result, path3, which consists of the 3-paths from nodes in I to nodes in II, via adjacencyI-match-adjacencyII arcs.   The entries in this matrix would therefore be equivalent to the number of 4-paths from a sceneI node (i) to itself, going via a sceneII node (j), if there were a match arc from j to i. Hence by clearing all those entries in path3 which do not have a match between them, the result represents the required weighted match graph.   This is then iteratively pruned by finding the entries in the matrix with the maximum (global) value, marking them as matched in a result bit-plane, and ignoring the rows and columns thus defined at the next repetition.

The implementation of the adjacency resolution algorithm is straightforward.   It consists of multiplying the adjacency matrix of the first scene with the sceneI to sceneII match matrix, followed by multiplication of the result by the adjacency matrix of the second scene.   The final product, path3, can then be taken as the actual weight matrix for resolution, without further processing, as described below.

Producing the matrix representing the adjacency relation consists of placing a binary one in the (i,j)th element of the matrix when region i is adjacent to region j.   (Since this is a symmetric relation, if the (i,j)th element is set, so must be the (j,i)th element).   To do so requires that (1) the regions be labelled in a repeatable manner and (2) the definition of adjacent be chosen.   Regions can be labelled using a variety of techniques, the most appropriate utilising the global propagation facilities of CLIP4 [23].

Superficially, the definition of the adjacency of two regions presents no problem: it can be expressed in terms of the sharing of a common boundary.   However, when viewed at the pixel level, the situation becomes more obscure. For instance, in Figure 7.10, regions a and b are in all cases adjacent, but what of b and d?   (Recall that regions are 4-connected and boundaries 8-connected.)

The solution chosen was to produce a method which gave results that matched intuitive expectations in simple cases, and to accept the definition so implied and its effect on borderline cases.

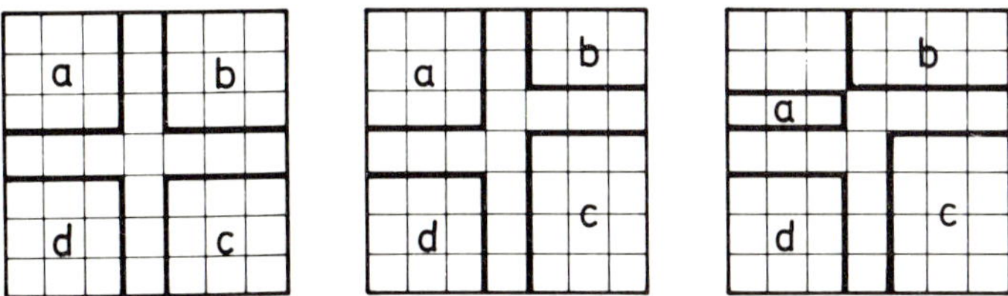

Figure 7.10   Adjacency

The definition decided upon was that all regions that could be connected to the region in question (the seed region), through the boundary pixels next to the seed, were adjacent to the seed. However, due to the 4-connected regions/8-connected boundaries distinction, a simple propagation through the boundary (requiring an 8-connected path) was not possible. The solution was to extract the boundary next to the seed, expand it (4-connected) and then OR the result with the original segmented scene. Thus the seed and its adjacent regions were cemented together by the expanded boundary so that a 4-connected propagation signal started from the seed's pixels would be able to extract a mask defining its adjacent regions.

4.4 Matrix Multiplication

Using a serial processor to implement matrix multiplication directly, requires three nested iteration loops, one for each of the i,j,k indices. Thus, in this basic form, the time to perform a matrix multiplication is $O(n^3)$, where n is a linear dimension of the matrices. Several methods of speeding up this process have been devised for serial processors in terms of reducing the n^3 asymptotic time complexity, the best currently being due to Strassen [54]. However, all the proposed enhanced methods rely on the property that the time taken to operate on a pair of values is independent of their position in the array. This condition does not hold in the case of a parallel processor of CLIP4's structure, and so such 'fast' methods are no improvement over a direct calculation that utilises the array's potentialities.

The method for multiplying two matrices A and B in parallel (based on Klette [55]), is to form matrix A'_k in which each column is equal to the kth column of A.

Similarly B'_k is formed of the kth row of B.   If the element-wise product of these two matrices is formed, the result will be a matrix C'_k whose elements are:

$$c'_{pq} = a'_{pq} * b'_{pq} = a_{pk} * b_{kq}$$

Thus C'_k corresponds to one term of the sum in equation (1). By iterating this for all values of k and adding each C'_k into an accumulator, the matrix product may be calculated. Since the pointwise multiplication and addition are performed in parallel, the time complexity becomes proportional to k reducing it from cubic to linear.

The basic implementation details of this algorithm are straightforward. An index plane has a marker bit set in it (corresponding to k) starting at the origin (i=j=0), which is progressively shifted along the major diagonal at each iteration, enabling the extraction of the kth row and column of the appropriate bit-stack. Once extracted, the column (row) is spread horizontally (vertically) with global propagation. The resulting two stacks are then pointwise multiplied and added into an accumulating stack. When the marker (k) has moved beyond the area in both arrays that contains significant information, as determined by a previously produced masking plane, the process stops.

This simple implementation requires too much bit-plane storage, however, to be executable wholly within CLIP4 (i.e. without using disc back-up). To overcome this, the fact that all stack arithmetic in CLIP4 is done bit-serially may be used. Since multiplication is accomplished using the shift-and-add principle, where in this case 'shifting' is simply a re-addressing of the bit-planes, the additions for the accumulator and the multiplication can be combined. This dispenses with the need for an intermediate stack holding the result of multiplication prior to accumulation.

Performing the matrix multiplication of two 5-plane stacks of 96 x 96 elements producing a 17-plane result took 4.1 seconds.

The time taken to perform the same operation on a PDP-11 was 67.8 seconds. This was using 16-bit integers since the arithmetic instructions do not operate on variable length operands.

Assignment process

Having produced a matrix representing the weighted matching graph (WMG), the final stage is to form a new binary match matrix based on this graph. This is

accomplished by using a masked global maximum routine in
which the points in a bit-stack corresponding to the elements marked in a candidates plane are compared and those
with the maximum value are marked in a result plane.

Applying this routine to the WMG matrix, with the
candidates plane set to those points that correspond to arcs
in the original match graph (i.e. the candidates plane is
initially equal to the original match matrix), marks those
WMG arcs with the highest weight. These, being considered
the most consistent matches, are ORed into a partial results
plane, and the nodes thus defined, along with all arcs connected to them, are removed from further consideration by
zeroing all rows and columns in the candidates plane containing a one in the results plane.

Repeating this sequence until all points have been
removed from the candidates plane leaves the resolved match
matrix in the results plane.

5. RESULTS

5.1 Predicted Position

The predicted position scheme for the resolution of
multiple matches was applied to each of the three sequences
using 3-plane segmentation and also to the 4-plane segmentation of the leaf sequence. This meant that the effects of
occlusion, velocity change and differing numbers of regions
could be examined.

The effect of velocity variation

It was found that slow accelerations had no adverse
effect on this method. However, abrupt changes in velocity, especially when linked with small, closely packed
regions, gave rise to resolution errors. The failures
generally resulted in incorrect multiple matches being
unresolved rather than erroneous removal of correct matches.

The condition under which predicted position fails,
appears to be when the steady motion condition (assumption
(1) of section 4.1) breaks down.

The effect of occlusion

By applying predicted position to the polygon sequence, all multiple matches were resolved correctly, with no false assignments. However, this result is somewhat fortuitous since the mild occlusion combined with the use of the bounding rectangles' centroids enabled the method to work under conditions that were expected to cause it to fail. The less controlled occlusion of the objects sequence, on the other hand, gave rise to several failures in resolving the correspondences.

Two main effects of occlusion can be identified. Firstly, the centroids of the regions' bounding rectangles, whilst having a certain insensitivity to occlusion, vary from their true positions. This causes their predicted positions to be in error. Secondly, as regions become totally occluded they cannot be tracked and so, when they re-appear, their velocities, and hence predicted positions, are unknown causing them to be unresolved for several frames.

Again these failures can be related to assumption (1): occlusion of a region alters its apparent velocity since only its visible part is considered in the determination of its centroid.

Timings

The predicted position resolution scheme was implemented in two parts: calculation of the regions' centroids, and the resolution stage proper. The centroids were produced as bit-column arrays and stored on disc. The resolution stage then used these plus the files of match matrices as inputs. The times quoted here (Table 7.1) do not include any of this I/O overhead in order to make clear the parallel processing aspects of the method.

Table 7.1   Times for predicted position resolution

Time (s)	Leaf3	Leaf4	Polygons	Objects
per frame	9.81	17.60	5.09	4.29
per multiple match	0.981	0.978	1.018	1.031

5.2 Adjacency Resolution

The same four sequences from which the predicted
position results were obtained, were used for adjacency
resolution analysis. However, since adjacency resolution
uses the contexts of regions, some properties of this technique related to variations in contexts can also be
described.

Velocity effects

The two leaf sequences, containing no occlusion, make
clear some of the effects of variations in velocity. It was
found that velocity changes alone had no effect on the adjacency resolution scheme, as was expected. This is because
assumption (3) (constancy of intra-object relationships) on
which the scheme is based, is independent of an object's
velocity. Figure 7.11 indicates the results obtained from
the leaf3 and leaf4 sequences.

Velocity effects can be responsible for variations in
the performance of adjacency resolution in circumstances
where one object's boundary passes so close to another's
that their boundary regions are adjacent. The changes in
contexts caused by the relative motions of the objects could
affect the resolution results, especially when large relative velocities are present. This can be seen as a break-
down in assumption (4). However, insofar as the touching
of object boundaries can be regarded as a degenerate type of
occlusion, this effect may be classed as a problem of occlusion.

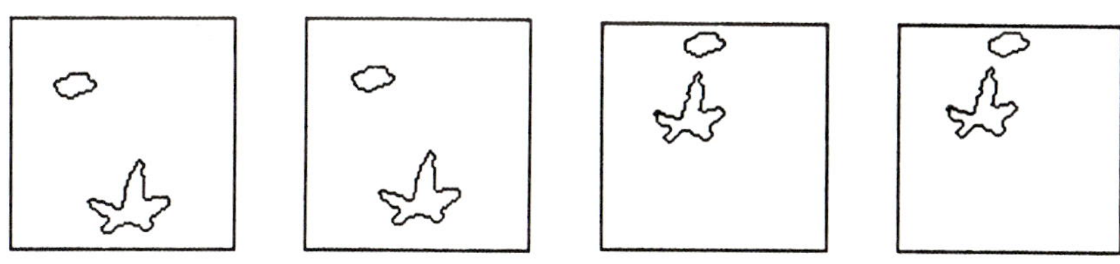

*Figure 7.11   Leaf3 and Leaf4 merged according to adja-
cency resolution.*

Context effects

The most common cause of failure in the adjacency resolution method was found to be when multiply-matched regions
have the same corresponding contexts, as depicted earlier in

Figure 7.6. The polygons sequence has a typical example of
this where the regions above and below the small triangle
have the same contexts and thus cannot be resolved
(Figure 7.12). It should be noted that, although the large
and small triangles are analysed as being part of the same
object by the merging routine, their common boundary points
cannot be deleted without apparently merging the two
multiply-matched regions in Figure 7.12. In spite of this
partial failure in resolution, erroneous matches between
these two regions and regions in the square object were
removed due to their differing contexts.

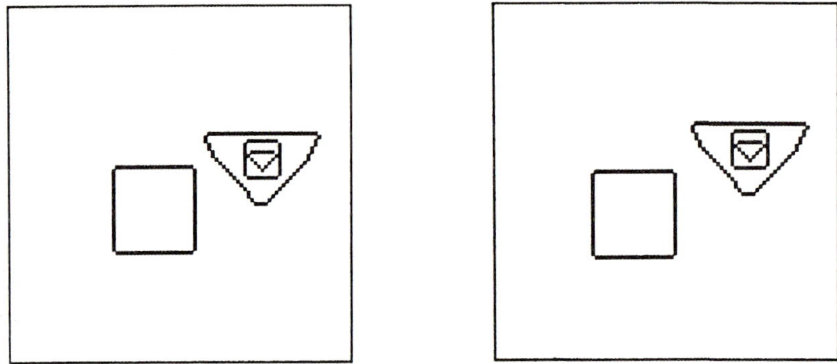

Figure 7.12   Partial merging after adjacency resolution.

Failures of this type are indirectly caused by the small
size of the regions' contexts: if a large number of regions
are in two surrounds, then the chances are greater that the
contexts will contain some different regions. This can be
regarded as occurring when assumption (2) is weak, that is
when objects are formed from few regions. The successful
resolution of the leaf sequences illustrates this since they
both show contexts with reasonable sizes for the multiply-
matched regions.

Occlusion effects

The adjacency resolution algorithm performs well on the
first few frames of the objects sequence, as can be seen in
the re-segmented images of Figure 7.13. When the objects
start to occlude however, some of the multiple matches are
not resolved due to the confusion produced by occlusion of
the regions' contexts. This gives rise to the unsuccessful
resolution of both the occluded regions (since their contexts are changing) and non-occluded regions, as seen in
frame 8, which are multiply matched to regions being changed

or created by occlusion. This sequence was, however, found
to be an improvement over the corresponding result using
predicted position.

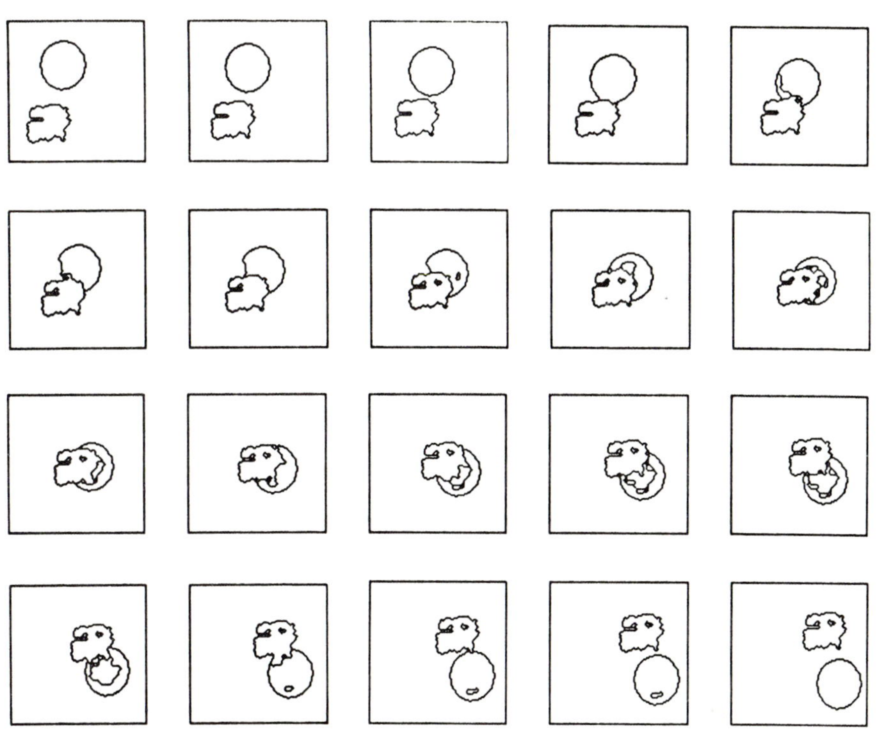

Figure 7.13   Objects merged according to adjacency
resolution.

Timings

Table 7.2 gives the execution times for the two procedures involved in adjacency resolution: calculation of
the adjacency relation matrix, and the resolution process
itself. The two processes were run separately using disc as
back-up image storage, consequently the quoted times do not
include this I/O overhead.

Table 7.2   Adjacency resolution timings

Time (ms)	Leaf3	Leaf4	Polygons	Objects
Adjacency relation calculation				
per frame	185	353	129	204
per region	13	15	13	14
Adjacency resolution				
per frame	970	1655	686	1054
per region	69	72	69	72
Total times				
per frame	1160	2010	820	1260
per region	83	87	82	86

These results show how both the calculation of the adjacency matrix and the resolution method are dependent upon the number of regions. For the production of the adjacency relation, this is due to the 'serial by region' nature of the processing.

The resolution times are approximately proportional to the number of regions since the main processing involved is matrix multiplication which, as shown earlier, has a linear time complexity.

5.3 Combined Results

It was found that the resolution failures of the adjacency method produced very few erroneous matches and these were of the type which cannot be avoided with this kind of analysis. The majority of failures were due to the algorithm only partially removing the false matches, while leaving the correct match arcs in place. This suggested that a subsequent pass of resolution by predicted position might be successful in removing the remaining multiple matches. In addition, applying adjacency resolution to the output of predicted position might also give improvements. The results are reported in full in [52].

Timings

Several interesting features can be deduced from Table
7.3 which shows the timing results for the two-part resolution processes. The first row in the table shows the time
per frame for an additional adjacency resolution stage after
processing the sequences by predicted position. Similarly,
the second row gives the times for an additional predicted
position resolution of the output from an initial adjacency
stage. The last two rows show the total times for each
combination.

Table 7.3   Times for combined methods

Time (s)	Leaf3	Leaf4	Polygons	Objects
additional adjacency	0.482	1.291	0.342	0.728
additional predicted position	0.249	0.495	2.144	0.379
total: pp-adj	10.291	18.891	5.433	5.015
total: adj-pp	1.219	2.150	2.830	1.433

Firstly, the times for adjacency resolution followed by
predicted position resolution (adj-pp) are seen to be significantly faster than pp-adj, for all the sequences tested.

Secondly, some measure of the overheads for each process
can be deduced since, for the polygons, being completely
resolved by predicted position, the additional adjacency
resolution stage had no further resolving to undertake.
Thus the 342 ms/frame may be interpreted as the time taken
by this stage to discover that no multiple matches exist.
Similarly, the leaves sequences show the overheads for
predicted position resolution, having been perfectly
resolved by adjacency. The dependence of predicted position, including its overhead, on number of regions is seen
by the comparison of the additional times for leaf3 and
leaf4.

Finally, and most significantly, the total time for the
polygons sequence to be processed by adj-pp is seen to be
2.83 s/frame compared with 5.09 s/frame for predicted
position alone. Since both pp alone and adj-pp produce
identical, perfectly resolved matches, this indicates a
strong preference for adjacency followed by predicted position rather than pp-adj or predicted position alone.

6. CONCLUSIONS

Resolution of multiple matches by means of the predicted positions of regions is an intuitively attractive idea. It has been seen that it is quite effective when the regions involved in a multiple match are widely separated and do not undergo sudden changes in direction. Under such circumstances, the differences in the predicted and actual positions for the correct match to a region are significantly less than those for the incorrect matches, resulting in successful resolution.

The failures of predicted position resolution occur when assumption (1) (that objects tend to move steadily) breaks down, as would be expected. However, the validity of assumption (2) (that objects have several regions) is also seen to contribute to errors since, if there are several small, closely grouped regions moving with the same velocity, the distances from predicted position to actual position for correct and incorrect matches are comparable, even for slowly changing velocities. The last point might be alleviated if a higher resolution input were available allowing greater precision in velocity, and hence position, calculation.

Adjacency context information was found to be effective in the removal of erroneous multiple matches in several situations in which predicted position failed. This is because the context of a region is independent of its velocity or acceleration for non-occluding objects. In fact, even for occluding objects, provided their relative velocities are small, assumption (4) allows useful contextual information to be gained from inter-object relationships.

Another advantage of adjacency resolution as part of an analysis system is that it is, in a sense, 'softer' than predicted position: when unsuccessful it tends to leave correct matches. This means that the opportunity exists for processing the resolved match matrix further to remove multiples with greater confidence that correct matches still remain.

The speed of adjacency resolution has been seen to be greater than three-dimensional matrix processing. This has meant, in addition, that the speed of adjacency resolution is almost independent of the number of regions, whereas predicted position's timing is proportional to the number of multiply-matched regions.

The main disadvantages of this method are that it works better with a large amount of context information and is consequently less efficient when there are few regions per object, whereas predicted position may be preferable when the number of regions is small. Secondly, changes in regions due to occlusion such as splitting, merging and disappearance, can confuse the algorithm giving wrong results.

The use of a combined adjacency and predicted position technique has been seen to be an improvement over either method on its own, especially when the order of use is adjacency followed by predicted position. The reasons for the preference of the combination in this order are twofold. Firstly, the speed is much higher than performing predicted position initially since the prior adjacency resolution will have removed a (possibly large) number of multiple matches which, given the need for predicted position to examine each multiple match individually rather than in parallel, reduces the burden on the slower process. Secondly, adjacency resolution's 'soft' nature allows a subsequent predicted position stage to work with match sets still containing the correct matches, whereas applying the two techniques in the opposite order gives rise to the possibility of adjacency having to work with erroneous data.

Thus a combination of adjacency resolution followed by predicted position resolution, removing multiple matches in stages, is considered to be a useful technique in the first stages of an image sequence analysis system. Further processing, probably involving some form of modelling, is required to overcome the remaining problems caused by velocity variation and occlusion.

CHAPTER   EIGHT

AUTOMATIC   SEGMENTATION

The long-term motivation behind much of the work described by David Reynolds in 'Automatic Generation of Image Segmentation Procedures for a Cellular Array' was the automation of some aspects of image processing and visual pattern recognition. The goal was to enable potential users of image processing to create solutions to tasks which require such processing, without them needing to acquire (or hire) extensive specialised knowledge of the field.

There are two main reasons why such automation might be important. Firstly, it is well known that the cost of computer software is rising rapidly and has passed the point where the software, rather than the hardware, is the determining factor in the cost of many applications. This problem is particularly severe in the field of image processing where the techniques are not yet well enough understood for the construction of a new vision system to be an everyday engineering task.

The second motivation for working on automated programming of image processing tasks was the hope that in doing so a clearer understanding of the problems involved would be gained, leading to a more formal knowledge of useful techniques. The task of organising image processing methods into an ordered array of tools with precise and consistent descriptions is an end in itself as well as a prerequisite for truly powerful automatic programming.

1. THE TASK AND ITS MOTIVATION

1.1 Introduction

The variety and complexity of problems in image processing, coupled with the lack of solid theoretical understanding of much of image analysis, means that all too often the solution to a minor problem requires an entire research project. This makes an image processing solution to many tasks uneconomic, even when the currently available hardware is adequate. Successful automation of even part of this process of software development should bring down the cost of image processing systems significantly. Prywes [56] has estimated that cost reductions of 80% over hand methods can be achieved by automatic programming in the field of commercial data processing. It is unlikely that such automatic programming could have a major impact on the development of large-scale systems intended for long-term applications (such as routine medical screening: blood analysis, cervical smear screening and karyotyping) for which the problems might be so difficult as to require extensive research, and for which the extent of the application justifies the development cost. It is feasible, however, for automatic programming to greatly reduce the cost of producing short-term applications modules where problems are not severe and the lifetime of the application does not justify extensive development work. Examples of such applications are automated assembly and automatic inspection on production lines producing varied small volume batches, or the analysis of small volumes of image data in scientific research.

Broadly speaking, there are two possible research strategies which might be adopted when tackling a problem of this nature: either restricted parts of the problem can be investigated in great depth and theories tested in isolation, or an attempt can be made immediately to build entire systems to perform the whole job, bearing in mind that for the same expenditure of resources no area will be covered in the same depth as if it were isolated. The conflict between these two strategies arises throughout the field of

artificial intelligence research, for example Tennant [57]
briefly discusses them in relation to natural language processing and Hayes [58] criticises the premature construction
of whole systems. The holistic approach of trying to build
an entire system has the advantage of forcing one to consider all the problems involved. It does, however, tend to
force compromises to be made during implementation (in the
interests of reasonable program speed or size) so that individual problems may not be properly solved, but just patched
over. The alternative approach, of investigating specialised subproblems in depth, is more likely to yield solid
theoretical understanding of those subproblems studied, but
it is hard to ensure that the crucial problems are being
studied, rather than interesting side issues.

   In this Chapter, a system is described and discussed,
which has resulted from a study of the feasibility of
automatic generation of classes of image processing programs. An attempt was made to construct a complete system
by restricting the image processing task to that of low-
level image segmentation.   Image segmentation is a particularly useful task to automate since, as well as being an end
in itself, it is also often a first step in larger tasks
such as object recognition and scene interpretation.   An
example of a direct application is the automatic analysis of
small volume scientific data where the problem might be to
locate objects in an image and to measure simple parameters
such as areas and perimeters.   This arises, for instance,
in the use of cell culturing as an assay technique in
experimental haematology where it is only necessary to count
the number (and sizes) of the cell colonies grown on a Petri
dish (see Chapter 5).   Similarly the analysis of sequences
of images of the organism <u>Naegleria gruberi</u> described by
Pass [59], required only low-level segmentation and simple
area measurement.

   So, to sum up, the short-term aim of the project to be
described was to create a program capable of generating
image segmentation procedures for a range of image classes.
It was also a pilot project for a much more ambitious system
capable of more general automatic programming in the domain
of image analysis.   The system that has resulted from this
project is called **Themis** and performs as described in the
following section.

1.2 Themis

The user presents Themis with a brief qualitative
description of the properties of the object class in the
images he wishes to process. This is done by answering a
short list of questions asked by Themis on the size, brightness and so on of the objects. He presents Themis with an
example of a raw image (via a camera or from a computer-
based file) and points to the relevant objects in the example using a cursor.

Themis generates a possible segmentation procedure which
should result in a binary image denoting the edges of the
objects in the raw image. It runs this procedure (using
the CLIP4 computer for all the image processing tasks) and
displays the resulting segmentation, overlaid on the original image. Themis then estimates the quality of the results
by assigning scores to each of the regions originally indicated by the user. It does so on the basis of how well the
region matches Themis's idea of the nature of the objects.
It also assigns a global score indicating the noisiness of
the result. If any of the scores have a low confidence,
Themis asks the user to supply his own scores. The user
can also override this automatic scoring system if necessary.

If the scores indicate that the result is satisfactory,
Themis halts its consideration of the current image, records
the raw image, the segmented image and various global
measurements for future reference, and awaits further examples to try. If the scores indicate that the result is not
satisfactory, Themis attempts to discover what caused the
failure and what to do about it. It then refines the
current segmentation procedure and tries the new version.
This 'trial – score – refine' cycle is repeated until a successful segmentation is achieved, Themis runs out of things
to try or the user abandons the program.

When a segmentation procedure has been generated that
produces acceptable results for the entire set of examples
available, Themis can then deliver:

1. The segmentation procedure as a subroutine in
    IPC source code.

2. A set of feature values for the objects which it
    measured during the processing. These features
    might perhaps be used as data for a simple
    region classifier as a later processing step.

Themis is not designed to be able to cope with arbitrary objects. The objects should be simple in that they can be segmented by local means (thresholding or edge finding for example). It is largely intended for use on structureless regions of uniform texture on a fairly uniform background. It can actually cope with some internal structure in the objects and spurious structure in the background provided that the objects have some well-defined bounding region. It does not, however, have any ability to deal with complex objects. For example, in Figure 8.1 the shaded regions would be regarded as separate objects and Themis could have no concept of the structure forming a single `object' (called a face). It could, however, deal with the object in Figure 8.2 by regarding it as a single region with holes. This same limitation means that images of three-dimensional objects with shadows are difficult for it because it has no understanding of the association between the shadows and the parts of the object that throw the shadows. Themis has a syntactic understanding of simple blob images but no understanding of complex structures nor of the semantics of images.

2. THEMIS IN OUTLINE

2.1 The Problem Specification

The purpose of the Themis project is to generate a low-level segmentation program for a class of images. The first problem which arises is how the task is to be specified to the program. The goal, to segment the image, is clearly fixed but, as has been seen, the meaning of this statement is largely subjective and varies between classes of images. It is necessary to specify to the system what the image class is and what objects are expected to be segmented out. Since, with the current state of theoretical understanding of segmentation, a formal specification is out of the question, a mixture of qualitative description, specification by examples and feedback on the results of trial runs is used.

The examples presented to the system are raw images of the target object class. It might seem reasonable for the user to hand-segment the images he presents as a detailed specification of the problem. This method of specification was not used for a number of reasons. Firstly, unless he

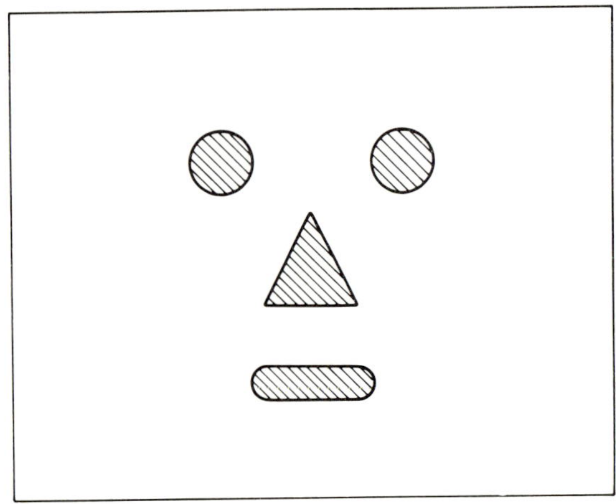

*Figure 8.1   Themis has no notion of complex objects –
this would be four separate regions (not a single face).*

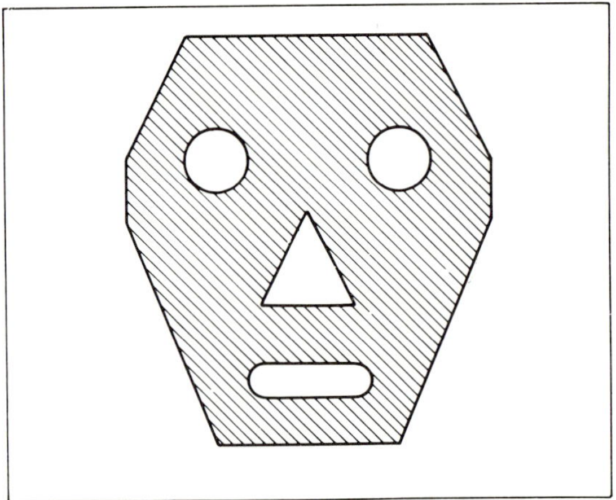

Figure 8.2   Regions with internal structure are admissible.

is to hand-segment all objects in the image, it will still
be necessary to verify attempted segmentations for their
effect on objects which are not pre-specified.   Secondly,
it is in the nature of the task that there is much flexibility in what is to be regarded as an acceptable version of a
segmentation.   As well as small-scale variations of one or
two pixels in the exact placement of a region boundary,
there may be internal structure in the objects or extraneous
structure in the rest of the image which a segmenter can
ignore or pick out without affecting its usefulness.
Specifying this flexibility other than by verification of
trials would be difficult and time-consuming.   Lastly, hand
segmentation can be very slow and tedious; it is pointless
if an automatic system is so unpleasant to use that people
prefer learning to write the programs themselves.

For these reasons the detailed specification of what the
segmenter should do is specified explicitly only by a loose
qualitative description of the target objects and implicitly
by feedback on the acceptability of trial segmentations.
This makes the task specification very simple for the user
since he only indicates example objects in the images and
then criticises trial results.   It seems to be much easier
to respond to specific examples of failures than to think
of, and describe, all aspects of a problem specification in
advance of concrete trials.   Another result of this approach
is that, as the program searches for a solution, the user
gains a much better feel for the nature of his images.
This clarifies his thinking on what he is really asking from
a segmenter and also shows what are the difficult problems
involved.   This means that even when Themis fails to
generate a completely acceptable segmenter it often gives
the user a good start on the problem. During the development of Themis it sometimes happened that a first run on a
new image set was not successful but the difficult aspects
of the problem were discovered.   A second run, with the
user penalising trials which failed to tackle these problem
areas, then tended to be more successful.

2.2 Overall Control Cycle

A simplified picture of the overall flow of control in
Themis is shown in Figure 8.3.   The first operation is to
acquire a qualitative description of the target objects.
This is used to guide the refinement of the segmentation

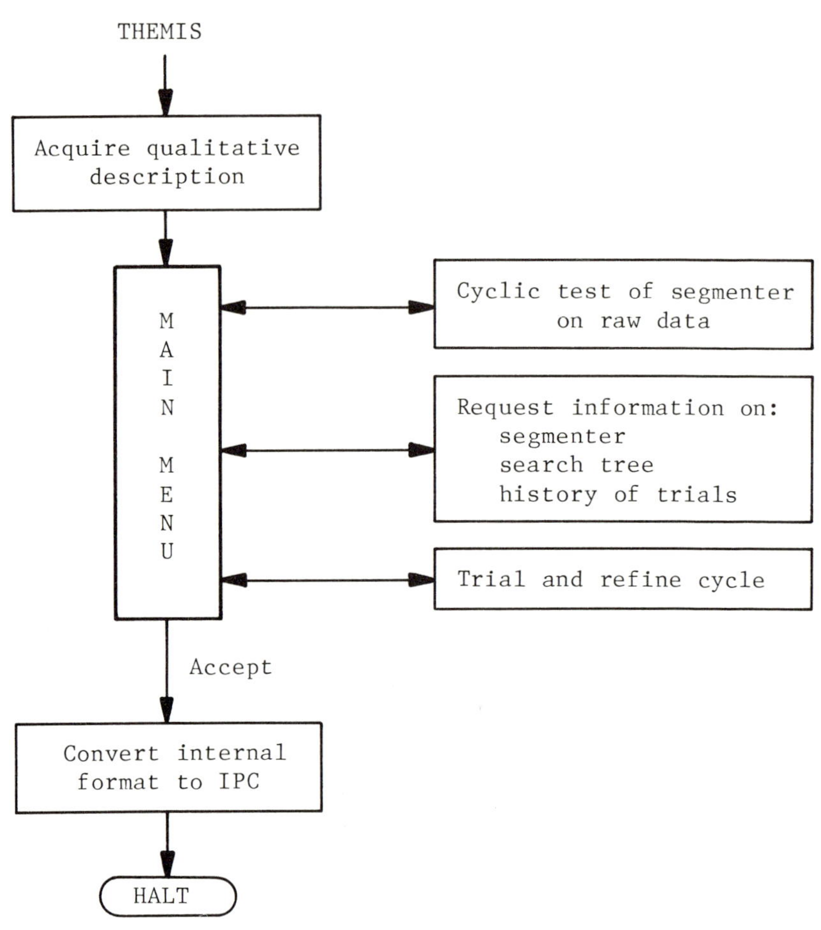

Figure 8.3   Overall Themis cycle.

procedure and for the automatic scoring of trial segmentations. As objects are successfully segmented these qualitative features are replaced by measured values.

The object description is acquired by Themis asking a series of questions on the objects and the user selecting from a short menu of options the description which best fits the objects. In the current system only a small number of very simple features are used for the testing of segmented objects and for each feature Themis asks how relevant the measure is. Relevance of a feature is specified on a scale of 0 to 10. If the feature is sufficiently relevant (relevance > 5) Themis presents a short menu of qualitative values to select from. An example of this interaction is (user's responses in bold font):

```
How relevant is `grey-level'?
    (integer range 0 to 10):  8
In what range does it fall?
    dark
    grey
    white
Option (abbreviations allowed):  dark
```

At this stage Themis also asks for a name for the object class. This name is used as the stem for the names of files in which the final outputs are to be placed; the qualitative object description is also placed in such a file. This allows the user to re-run Themis on an object class he has tried before and just reload his previous description.

Having read in or reloaded the object description, Themis then moves on to the main command loop. For simplicity the commands are entered as menu options. The usual action at this point is to try a new image, but, whenever an image has been successfully segmented, Themis returns to this main command menu where the other commands available are for use after one or more trials have been completed. The commands allow the user to see what the current segmentation procedure is, to review all the trials so far and to test the current segmenter on repeated camera frames to check for stability. Additional options allow examination of the internal workings of Themis. In particular a **comment level** variable can be modified to increase or decrease the level of running commentary Themis produces. At one extreme it only asks the user questions and volunteers nothing, at the other extreme debugging information is provided. At intermediate levels of comment Themis prints a description of why it takes a given action.

When all example images have been tried, Themis finishes by converting its internal description of the segmenter into a more usable form. It generates a simplified translation of the segmenter as a text file; a full subroutine (in IPC source code) to implement the segmenter, including store allocation code; and a file of measured values of the simple features used in the initial qualitative description.

2.3 Overall Refinement Cycle

Presenting a new image

The actual work of Themis is initiated by the user selecting the 'try a new image' command. The sequence of actions which follows is shown in Figure 8.4. The first action is to take in a new image to be segmented. This image can be taken from a camera, a UNIX file or from the **history list** of images already tried. This last option is important because one major failure of Themis is that the refinements which are needed to accommodate a new image are not guaranteed to preserve previous behaviour. This problem only arises when images exhibit a large variation and Themis can often generate a successful segmenter despite its short-sighted approach. However, it is important for the user to be easily able to retest previous examples. The new image is stored in CLIP4 which is used for all image manipulation. Raw camera images are 6 bits long but disc-based images can be of any length, limited only by the 32 D-levels in CLIP4.

The user is next asked to use a cursor to indicate relevant objects in the image and a **background** region. It is assumed that the image consists of one or more objects (target objects or otherwise) on a fairly uniform (though possibly textured) background. Indicating an example of the background region enables Themis to contrast its local properties with those of the objects before it has generated an object/background separation. The co-ordinates of each region marked by the user are recorded in an appropriate data structure. Later, if the corresponding region of a segmentation is required, a point is generated in CLIP4 at that co-ordinate and then used as a label.

For each indicated point, Themis asks whether the region is a target object or an object of a different type that it should be aware of. Themis is expecting to score trial segmentations only on the basis of the target objects indicated (the rest of the image counts in the global score) but

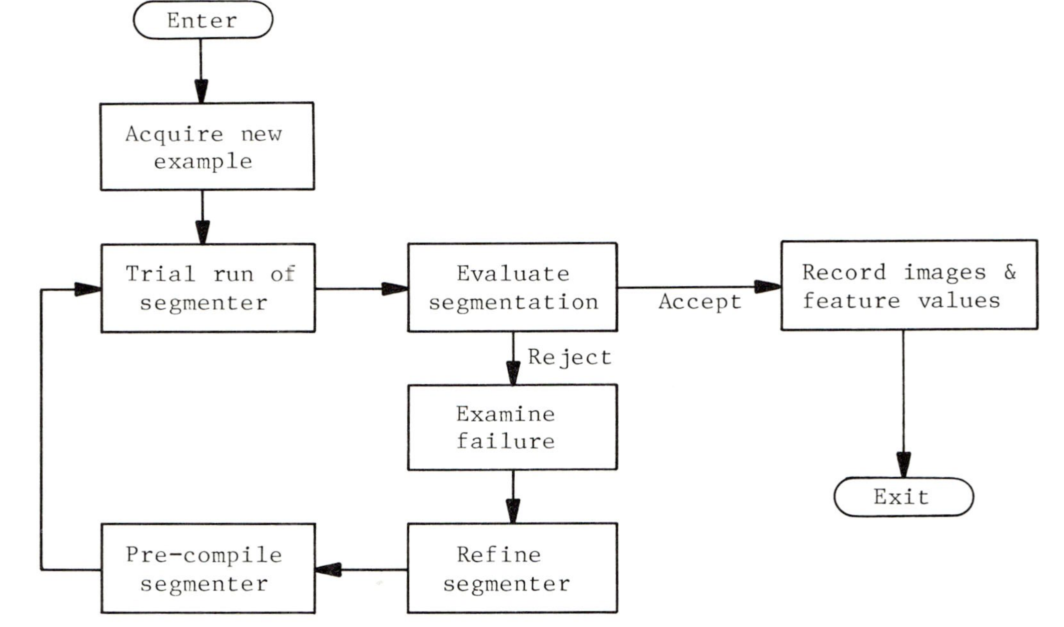

Figure 8.4   Themis - main refinement cycle.

if there are other objects present which are irrelevant to
the segmentation it may be desirable to point them out.   A
typical example is where there is a piece of extraneous
material close to a target object.   If the extraneous
object is indicated as well as the target then Themis can
automatically tell if a segmenter has accidentally joined
these regions together and act accordingly.

Trial segmentation

Having stored the example image and noted the location
markers for the objects, Themis runs its current version of
the segmenter. The segmentation which results is stored as
a binary edge map in CLIP4.   The edges are stored as
8-connected 1 pixels.   Thus the regions are regarded as
4-connected (see [29] for a description of elementary digital geometry).   Note that edges are denoted by pixels
rather than by interpixel boundaries. For example, a segmentation of Figure 8.5(a) might be that shown in Figure
8.5(b).   The region of a segmentation corresponding to a
given object marker is trivially found by a 4-connected
labelling operation, which is a single CLIP4 machine
instruction.

```
5   6   5   6   5   6   5            .   .   .   .   .   .   .

7   5  12  13  15   6   5            .   .   1   1   1   .   .

5  14  20  21  20  11   5            .   1   .   .   .   1   .

5  12  21  22  21  10   5            .   1   .   .   .   1   .

5   6  12  20  10   6   5            .   .   1   .   1   .   .

4   5   5  10   5   5   5            .   .   .   1   .   .   .

            (a)                               (b)
```

*Figure 8.5   Format of segmented images: (a) grey
levels of a simple blob image; (b) segmentation edge map of
(a).*

The segmented image is presented to the user on the
CLIP4 display, overlaid on top of the raw image (Figure
8.6(b)).   This enables him to see how the result
corresponds to the original data (Figure 8.6(a)).

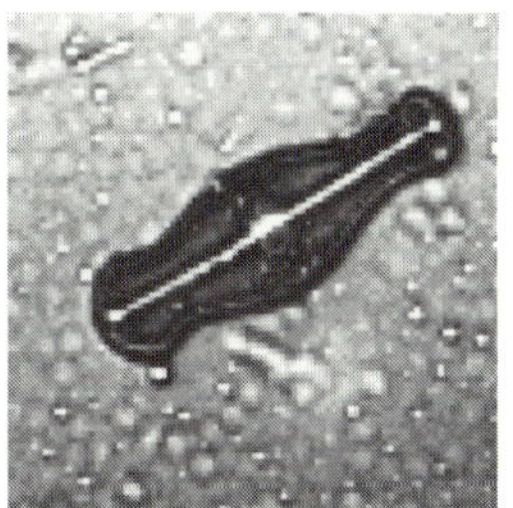

(a)

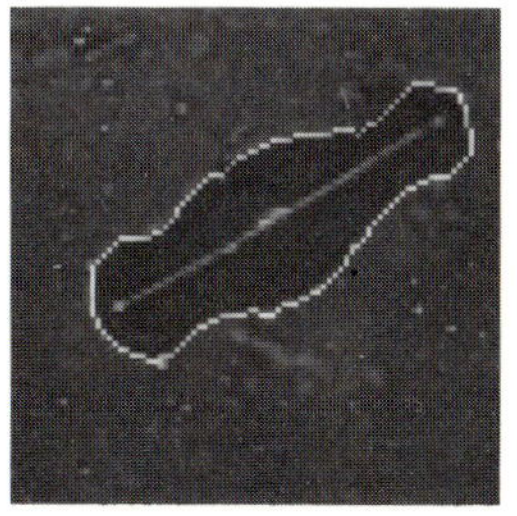

(b)

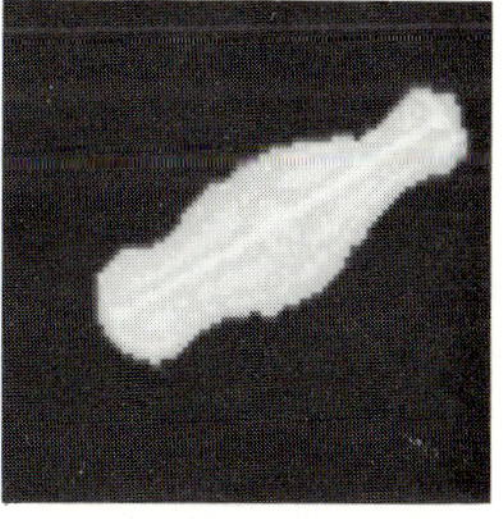

(c)

*Figure 8.6   Themis display formats: (a) raw image;
(b) image overlaid with segmentation edge map; (c) as (b)
with one object region displayed.*

Scoring the segmentation

Themis now estimates scores for each indicated region and for the overall segmentation. These scores take values in the range [-100, 100]. Positive scores indicate an acceptable segmentation and negative scores indicate failures. The magnitude of the score indicates how sure Themis is that the segmentation is good or bad. Each score whose magnitude is too low is confirmed with the user. This is done by displaying the offending region (see Figure 8.6(c)) and asking the user for his score for that region. The scores are presented to the user as shown in Table 8.1. The 'Last score' column gives the score assigned to that region on the previous trial and is a useful guide to the user when assigning scores by hand. It is important that the scores are consistent from cycle to cycle, otherwise Themis may incorrectly assume that a refinement has caused some improvement or degradation.

Table 8.1   Example score table - entries in bold are provided by the user.

Region	Last score	Estimated score	New score
1	-100	0	: **80**
2	-100	0	: **-20**
3	80	52	52
Global	50	47	47

Before Themis finally accepts these region and global scores, the user is given a chance to either confirm or change them. It is thus possible to override the automatic scoring system if required.

If all the regions, and the global score, are positive then the segmentation is assumed to be satisfactory. Themis verifies with the user that this is indeed the case; if it is not, Themis assumes that one of its automatically generated (and as yet unverified) scores is in error and asks the user to correct it. If the user verifies that the segmentation of the image is acceptable, Themis measures the feature values of all the target objects present and adds them to its record of object descriptions. It then notes the image and its verified segmentation on a **history list** for future reference. Thus, as each image is processed, Themis accumulates more data on the target objects which will guide its actions on the next example.

If the segmentation is not acceptable, Themis examines the nature of its failure, **refines** the segmentation procedure accordingly and compiles the new procedure into a form that can be efficiently executed. The new procedure is run on the current example image and the result is displayed and scored as before. This cycle repeats until a successful segmentation is achieved.

The first stage in examining a failure is to run a sequence of IPC subroutines which measure specific features of the segmentation (and raw image), such as the number of small regions in the segmentation, or the number of region markers which are labelling identical 4-connected regions of the segmentation. These segmentation measures will be called **predicates** and, together with the object feature values (measured from previous examples or obtained from the initial dialogue) and the list of failed regions, are the only information on the trial segmentation that the refiner uses. The predicate evaluations are the only direct image measurements carried out during a refinement cycle. For simplicity of implementation they are all recalculated each cycle, rather than only on demand.

Refining the segmenter

The deduction, from these direct measurements, of a suitable refinement to the segmenter, and the implementation of the selected change, are carried out by the **refiner**. The refiner is implemented as a set of 150 **rules**, each of which encapsulates a separate piece of knowledge used in the refinement process. Encoding the refiner in this way greatly eased the process of developing Themis and provides an easy means of recording the deductions used in the selection of a segmenter refinement. Each rule has a number and an associated block of text, which was provided when the rule was written. Each time a rule is **fired** (i.e. is tested and found to apply to the current data) the textual explanation of the rule can be printed and the rule number can be recorded on a **transaction** file for future analysis. The running commentary provided by the text of the fired rules can be enabled or disabled by means of the comment level variable mentioned earlier.

This rule structure has the flavour of a rule-based production system [60] but is not such a system in its current implementation. In a true production system an entire set of available rules is tested for applicability each inference cycle and an arbitrator selects a relevant rule to fire. The Themis refiner tests each rule in sequence and fires any applicable rule immediately. The

sequence can be varied at run-time, according to context, by
the rule equivalent of a branch operation. This control
strategy was found to be adequate for the task at hand and
allows the programmer to build in useful paths through the
rule base, leading to very fast interpretation of rules.
The penalty paid for the building in of this control information was some loss of ease of incremental extension to the
rule base. New rules have to be slotted into the correct
place in the rule base and not just tagged on the end. The
patterns and actions in the rules can be arbitrary C functions if need be, so that each rule tends to encode quite a
large amount of knowledge. This keeps the volume of running
commentary to manageable levels. In effect, the Themis
refiner is coded in a tailored applications language,
designed for this one job, rather than in a general rule-
based production system format.

These rules use the predicate measurements, and context
information from previous refiner cycles, to hypothesise a
possible cure for the failed segmenter. The resulting
hypothesis suggests a class of changes such as: 'add more
low-pass filtering before the threshold step'. Then a different subset of the rules implements a specific modification to the segmenter along the suggested lines. In the
above case a 5 x 5 mean filter might be selected and slotted
into place in the segmentation procedure. At any one time
the segmenter is represented internally as a partially
expanded AND/OR tree of possible segmentation strategies
called the **Sp-tree**. The changes made by the refiner
correspond to selecting alternative descendants of OR-nodes,
or modifying parameters attached to leaf nodes. Thus the
above modification would consist of replacing a dummy
descendant of the low-pass filter step by the 'mean filter'
leaf, and setting the leaf's parameter to correspond to
5 x 5.

This tree representation of the segmenter is very convenient for the refiner but contains much information which
is irrelevant to the current segmentation procedure. Thus
after a refinement has been made to the Sp-tree, it is
translated into a compact series of calls to IPC subroutines
which perform the actual segmentation. It is this translation of the Sp-tree which is run by Themis to generate a new
segmentation.

The full cycle of operations on Figure 8.4 have now been
completed from trial run, through scoring, failure measurement, refinement and translation and a new trial can
commence.

3. RESULTS

It is difficult to give a quantitative assessment of Themis for two reasons. Firstly, since the definition of segmentation is somewhat subjective, it is not possible to give an objective measurement of the quality of the synthesised programs. Secondly, Themis is intended to work over a range of image types but, as a theoretical description of the task is lacking, it is difficult to extrapolate the system's behaviour on unknown image types from a limited number of trials.

However, on the examples tried (approximately 70 images from 15 image classes which were selected as being within the scope of Themis's segmentation primitives), Themis managed to produce an acceptable segmenter for each class. In most cases the system appeared to be limited by the simplicity of the segmentation primitives available rather than by its ability to search for a solution. In particular, images of side-lit industrial parts, with shadows and highlights, were essentially beyond the system because it has no understanding of the relationships between, say, the shadows and the object that casts them. So, typically, Themis would attempt to segment the shadows as an extra class of objects, along with the part itself. The results of a number of test runs are now presented.

3.1 The Cross Images

The first examples are sufficiently simple to serve to introduce some of the system's basic patterns of behaviour. They were not amongst the test data set which was used for the initial development of Themis and so constitute a proper test of the system's capabilities. Each image consists of a two-tone picture of a cross with additive noise of known statistics. The segmentation of these images was suggested by Professor K Preston Jr as part of a standard task for comparing the performance of various image processing machines.

The first image, **cross1** (Figure 8.7(a)), consists of a cross with arms of grey-level 32 and a central core of grey-level 64, with additive gaussian noise of mean grey-level 128 and variance 32. Note that Themis only sees the raw image and is not provided with this statistical description. The trial Themis run went as follows:

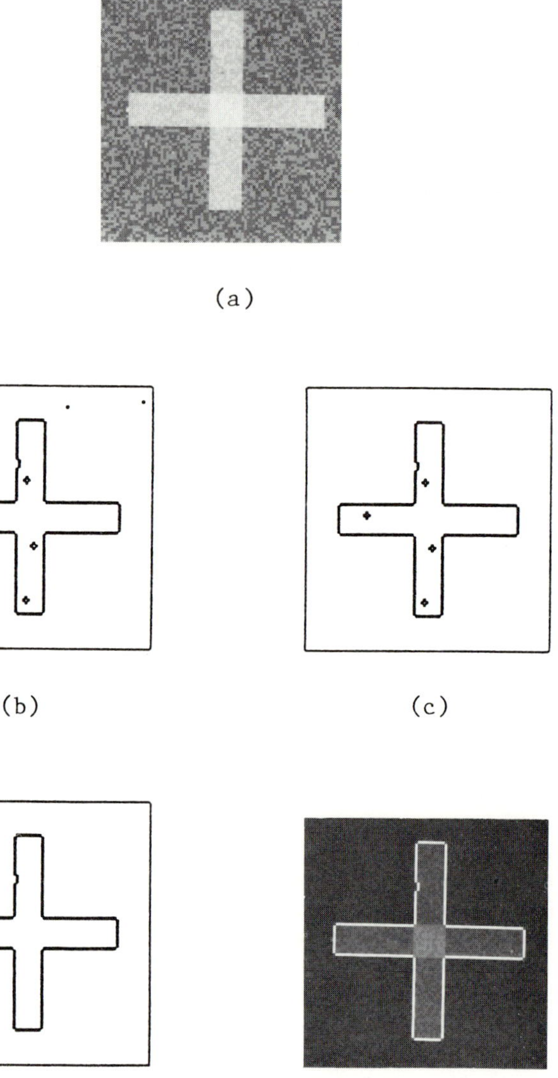

(a)

(b)   (c)

(d)   (e)

Figure 8.7   The segmentation of cross1:   (a) raw image;
(b) initial trial segmentation;   (c) clean_active_lines
added;   (d) merge_small_regions;   (e) overlay of result.

1. Themis ran the initial segmenter (**auto_threshold**
    only) to generate a reasonable result (Figure
    8.7(b)). The scoring system delivered a score
    of 6 for the object region, indicating a barely
    acceptable segmentation, and 31 for the global
    score, indicating some noise but nothing seri-
    ous. In fact the global noise (some single
    pixel background points and single pixel holes
    in the region) was not acceptable so the
    automatic score was overridden and set to -10.
    The refiner then deduced failure modes of
    spurious lines and **minor noise** and a context of
    global noise only. Since the region had been
    accepted and the problems seemed minor, post-
    processing was chosen and the highest hypothesis
    of adding **clean_active_lines** was selected and
    implemented.

2. The new segmenter generated a clean segmentation
    of the cross with some small holes left (Figure
    8.7(c)). Again the automatic scoring system
    accepted the region but with a low score and
    indicated minor acceptable global noise but was
    overridden. The refiner deduced the failure
    mode of **minor scale noise.** The hypothesis of
    adding line deletion had succeeded in curing its
    target fault without causing other problems so
    it was rated as complete. The next hypothesis
    selected was to add **merge_small_regions.**

3. The new segmenter generated an acceptable result
    (Figure 8.7(d)). The final program translated
    to:

```
auto_threshold
merge_small_regions (radius 1)
clean_active_lines
```

    which ran in 0.2 s.

Note that by accepting the region segmentation in the
first trial the user forced Themis to regard the small holes
as minor global noise. If the region score is overridden
and set negative then Themis regards the problem as too
serious to use postprocessing and instead adds mean filtering of radius 1 (3 x 3). This cures the noise problems in
one go (Figure 8.8(b)).

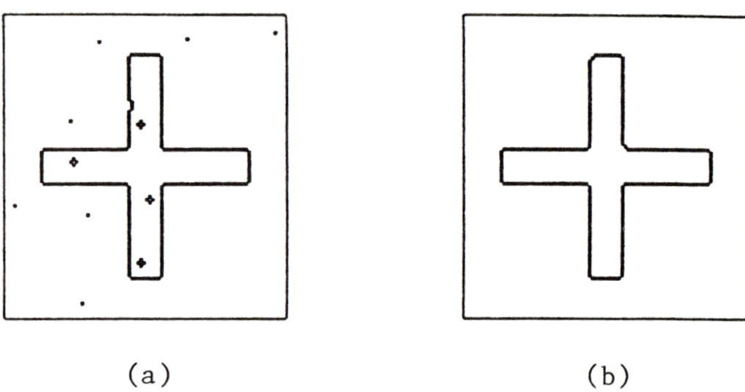

(a)   (b)

*Figure 8.8   Alternative segmentation of cross1:
(a) initial trial;   (b) mean filter added.*

The second cross image, **cross2** (Figure 8.9(a)), is
similar to cross1 except that the additive noise is
Laplacian with variance 1024, thus resulting in a much lower
signal-to-noise ratio.   The Themis run on this image was:

1.   The initial segmenter (**auto_threshold**) was a
     complete failure (Figure 8.9(b)).   The initial
     hypothesis selected was to add low-pass
     filtering.   The first filter tried was a mean
     filter of minimum size (radius 1).

2.   The new segmenter generated an improved result
     (Figure 8.9(c)) so the hypothesis was retained
     and Themis began a hill-climbing trial on the
     size of the mean filter.

3.   The next segmenter, with the mean filter at
     radius 2, showed more improvement.

4.   When the filter size reached radius 3 the seg-
     mentation started to degrade again (Figure
     8.9(d)) so the parameter was backed up to radius
     2.   Since low-pass filtering still seemed to be
     a useful technique, the next alternative filter
     was selected for trial.   Thus the mean filter

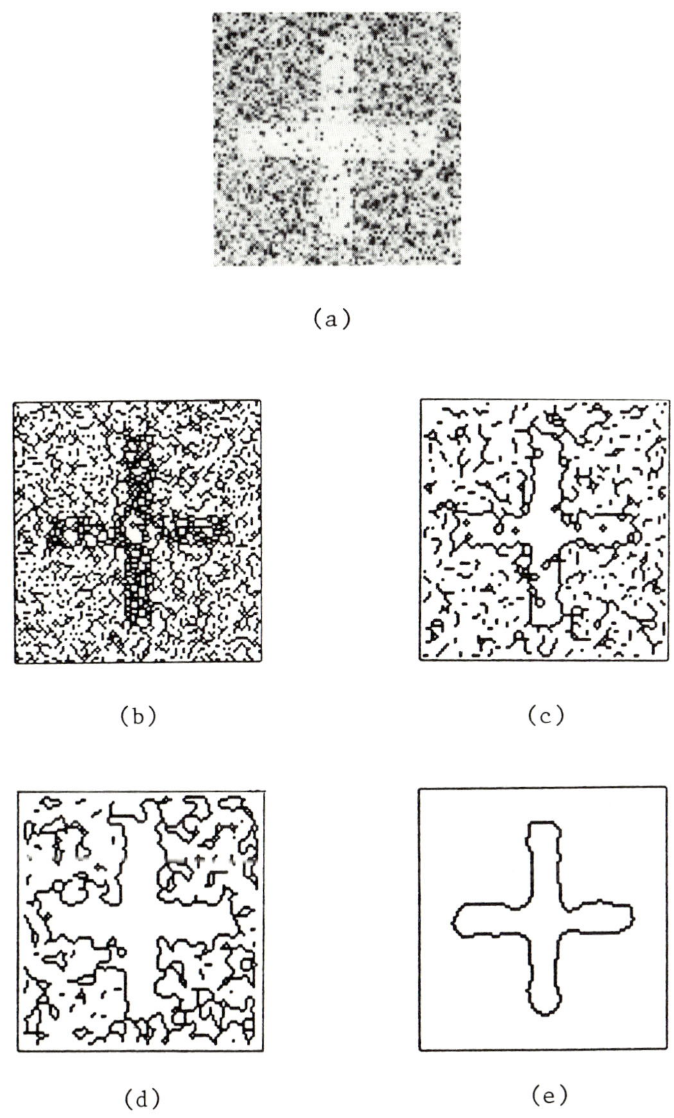

Figure 8.9   The segmentation of cross2:   (a) raw image;
(b) initial trial;   (c) with mean filter, radius 1;
(d) with mean filter, radius 3;   (e) with median filter,
radius 2.

was replaced by a median filter of radius 2 (the size of 2 was chosen because it had been a locally optimal setting for the first filter tried).

5. The result was acceptable, Figure 8.9(e), and the final program translated to:

```
median_filter (radius 2)
auto_threshold
```

This program ran at 3.2 s per image and is the slowest yet generated by Themis. The speed is so low because the work space needed to perform an iterated 5 x 5 median filter on an 8-bit image exceeds the storage space of CLIP4 making disc swapping necessary.

3.2 The Plane Images

These images consist of noisy pictures of small white objects on a highly textured background (Figure 8.10(a)). The background exhibits three features:

1. Uniform, short-range, texturing (of mid-range grey level).

2. Large dark areas of reduced texturing.

3. Small, bright, line noise caused by video-tape dropouts.

The Themis run went as follows:

1. The initial segmenter was tried and resulted in a barely acceptable region with serious global noise (Figure 8.10(b)). The scope of the noise, its uneven distribution and the small size of the target object led to the hypothesis **increase high-pass filtering.** The initial filter size was set to radius 5, just above the estimated object size.

2. The new segmenter resulted in an improved region but still with serious global noise, though the noise was more evenly spread (Figure 8.10(c)). High-pass filtering having had the desired effect, the next hypothesis was to add low-pass

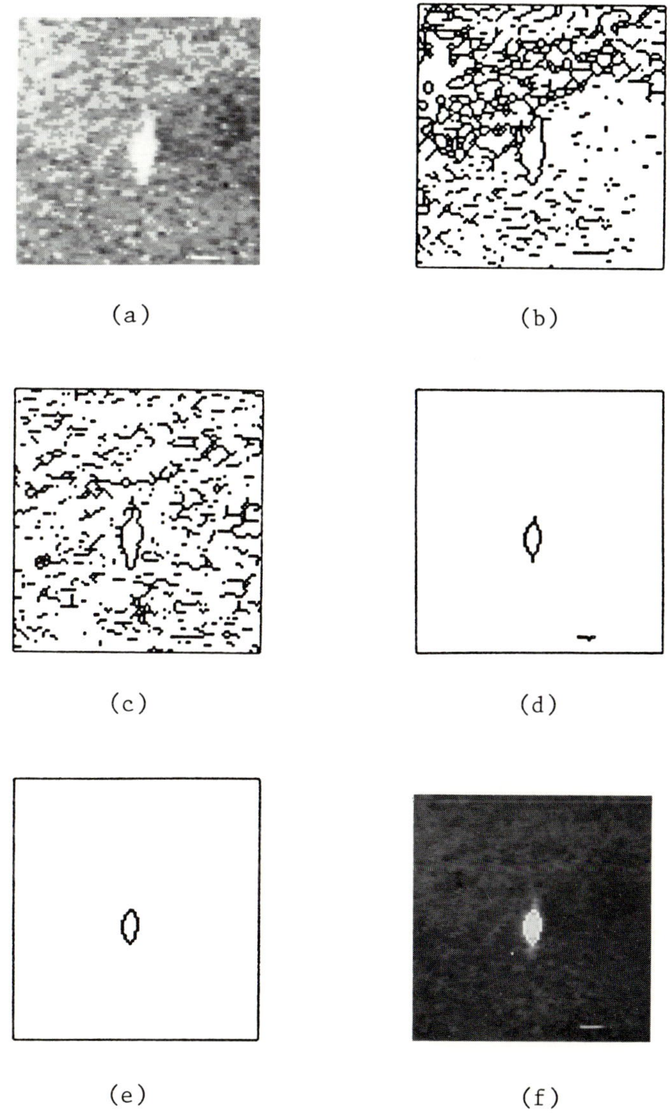

Figure 8.10   The segmentation of planes(0):   (a) raw image;   (b) initial trial;   (c) high-pass filter added; (d) low-pass filter added;   (e) clean_active_lines added; (f) overlay of result.

filtering to reduce global noise. Since the objects were small the filter size was set to just below the object size (radius 2).

3. The result of the extra filtering was a good segmentation with the exception of a spurious line associated with a dropout noise region (Figure 8.10(d)). This being a relatively minor global problem, the postprocessing step **clean_active_lines** was selected.

4. The resulting program generated an acceptable segmentation (Figures 8.10(e) and 8.10(f)).

5. The program was tested on five other images from the data set and worked on all but one. The segmenter failed on this image (Figure 8.11(a)) because a large dark area distorted the grey-level histogram and prevented the auto-thresholder from locating the very small object population (Figure 8.11(b)). After this failure Themis chose to try alternative thresholders.

6. The second choice of thresholding algorithm was the histogram-based thresholding. The object population was too small to yield any valleys in the grey-level histogram so this thresholder returned a completely blank segmentation. This was detected by the scoring system which declared the trial a disaster and skipped the user verification stage. The refiner includes rules concerning this behaviour of the histogram thresholder. One of these rules caused the useless node to be pruned from the Sp-tree and immediately tried the next choice of thresholder, which was **dual_auto_white.**

7. The resulting program generated an acceptable segmentation (Figure 8.11(c)) on this and all other images in the data set and ran at 1.2 s per image.

3.3 The Motion Images

The next set of examples is taken from a data set generated by Alan Wood for use in studying motion analysis of images (see Chapter 7). Each frame consists of three separate grey objects (obtained from a real scene)

(a)

(b)

(c)

Figure 8.11 The segmentation of planes(3): (a) raw image; (b) result using current segmenter; (c) thresholder changed to dual_auto_white.

artificially superimposed, one on another, to generate a moving sequence of occluding objects (Figure 8.12(a)). The main difficulty in this segmentation task is to separate the occluding objects at their boundary, despite its relatively low contrast.

1. The initial trial of the auto-thresholder (Figure 8.12(b)) segmented the object groups perfectly but failed to split the overlapping objects. Themis first selected high-pass filtering to enhance the object differences.

2. The high-pass filtering had no effect since these artificial images have a noise-free background. The high-pass filter was thus pruned from the Sp-tree and the alternative hypothesis of testing other thresholding techniques was selected. The next choice of thresholder was the histogram-based thresholder.

3. The histogram thresholder successfully segmented the overlapping regions (Figure 8.12(c)) but caused one object to be split into subparts. The successful object separation caused the refiner to suspend the examination of alternative thresholders and examine the new form of failure. The hypothesised cure was then to increase the node **low-pass filter** so a radius 1 mean filter was added.

4. The mean filter caused a slight improvement so the refiner increased its radius.

5. The radius 2 filter proved too much and the segmentation quality fell again. So the refiner backed up to radius 1 and switched to trying median, rather than mean, filtering.

6. The median filter generated a very clean segmentation (Figure 8.12(d)) except for a small noise region and some spurious corner effects (characteristic of the median filter used). Themis therefore added postprocessing, first **clean_active_lines** and then **merge_small_regions** to delete these problems.

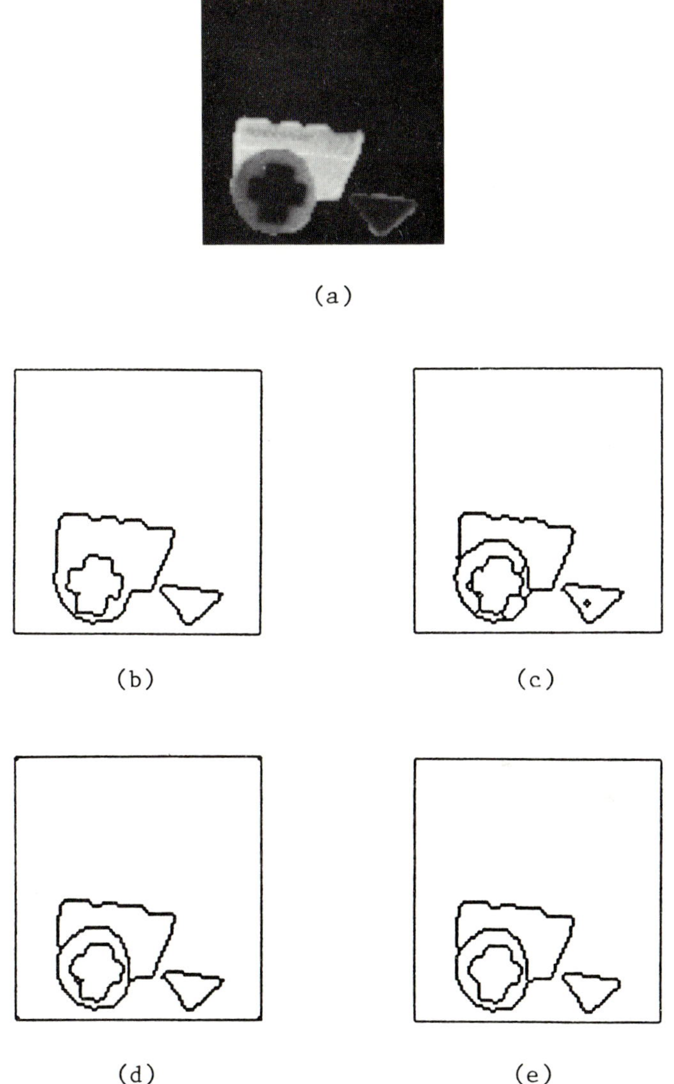

Figure 8.12   The segmentation of motion: (a) raw image; (b) initial trial; (c) switch to histogram-based thresholding; (d) median filter added; (e) postprocessing added.

7. The final result (Figure 8.12(e)) was accepted
    and the resulting program of

```
median_filter (radius 1)
histogram_threshold
merge_small_regions (radius 1)
clean_active_lines
```

worked adequately on all 32 images in the data
set and took 0.75 s per image.

3.4 The Diatom_c Images

The **diatom_c** data set illustrates the use of edge segmentation methods by Themis.   It consists of six images of
a particular diatom (Figure 8.13(a)).   This image is too
difficult for the simple thresholding methods available to
Themis because the background lighting is not uniform; any
threshold merges the object with the darker parts of the
textured background.   The object is too large in relation
to the array size for high-pass filtering to be applied to
correct for this non-uniformity. The trial went as follows:

1. After   17   refiner   cycles   Themis   rejected
    threshold-based   segmentation   and   moved   on   to
    consider   edge-based   methods.   The   initial
    edge-based method was that of expand/shrink edge
    detection,   followed   by   auto-thresholding   and
    thinning.   The result (Figure 8.13(b)) was a
    broken   object   border   with   some   global   noise
    lines.   The refiner elected to examine alterna-
    tive edge-thresholding methods.   The next choice
    was two-dimensional histogram-based edge thres-
    holding.

2. The histogram-based thresholder produced a sig-
    nificantly worse result and was pruned from the
    Sp-tree.   The next choice of edge thresholder
    was the watershed ridge-finding algorithm [61].

3. This   combination   resulted   in   a   good   region
    border   (Figure   8.13(c))   but   lots   of   global
    noise.   The refiner therefore decided to add
    median filtering as a preprocessing step.

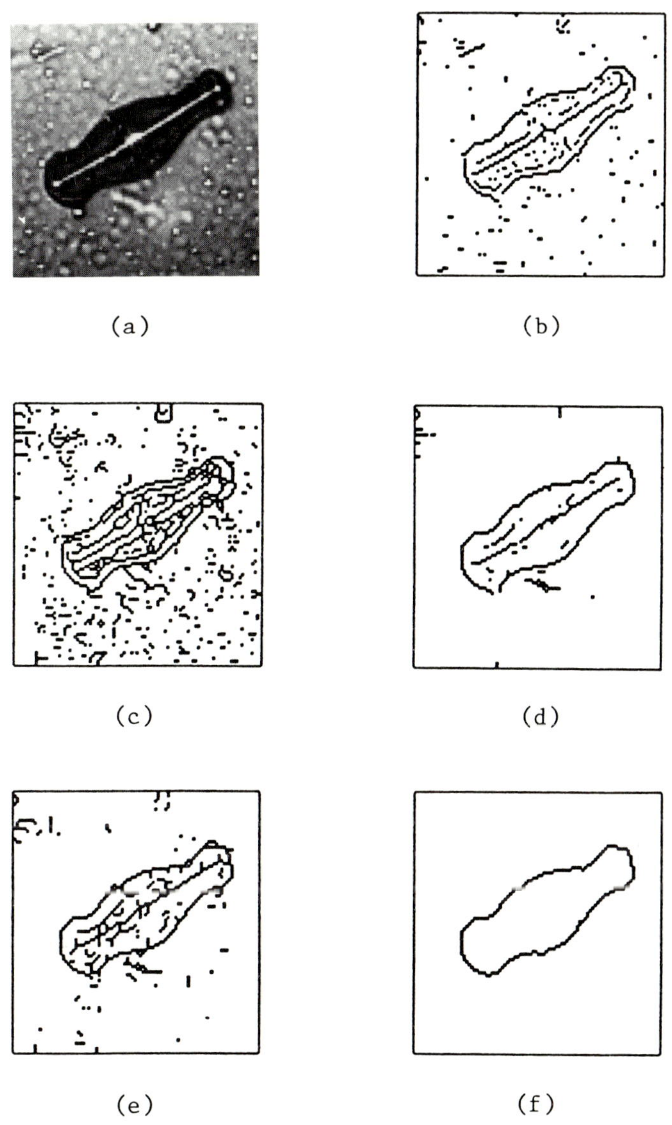

Figure 8.13   The segmentation of diatom-c: (a) raw image;   (b) first edge trial;   (c) switch to watershed ridge-finding edge thresholder;   (d) median filter added; (e) watershed cutoff reduced;   (f) postprocessing added.

4. This resulted in a cleaner edge map (Figure
    8.13(d)) but a small break in the object border.
    The refiner then chose to lower the threshold on
    the watershed ridge heights to allow the algo-
    rithm to detect weaker edges.

5. After two cycles of watershed tuning the result
    of Figure 8.13(e) was obtained. The region
    border was thus intact and the refiner moved on
    to add postprocessing. In two more cycles
    clean_active_lines and **merge_small_regions** were
    added.

6. The final result was acceptable (Figure
    8.13(f)). The program translated to:

> median_filter (radius 1)
> expand/shrink grey edge detector
> watershed (ridge height threshold of 3)
> thin_edge_lines
> merge_small_regions (radius 1)
> clean_active_lines

This segmenter worked on all six examples and ran at 2.5 s
per image (due to some disc swapping).

4. CONCLUSION

An interactive program has been described which automatically generates low-level image segmentation algorithms.
This program has produced acceptable segmentation algorithms
for a wide range of simple images. It is primarily limited
by the low-level nature of the segmentation primitives it
employs, rather than by the sophistication of the algorithm
synthesis. Even in cases where a completely successful
segmentation is not acheived, the user may be able to derive
useful insights into the nature of his target images by use
of the program.

CHAPTER   NINE

FURTHER   DEVELOPMENTS

*CLIP4 was one of the first of the now growing family of
parallel processor array systems.   It has proved to be of
considerable value in practical applications, algorithm
development and investigations of the type and range of
problems to which such array processors are usefully applicable.   However, after several years of operation, it
became apparent that two important developments of the
system, one practical and one theoretical, should be investigated.   These developments are addressed in this final
Chapter.*

*The first development concerns the practical matter of
pixel resolution.   Although the application of VLSI technology can, at present, permit the integration of about
sixty-four CLIP4-like processors on a single chip,
such improvements are overwhelmed by the increase in pixel
resolution required by some applications.   Many require
512 x 512 resolution, and some use images several thousand
pixels square.   No projected improvements in technology in
the near future will permit practical systems to be constructed with millions of processors.*

*In an early Chapter of his thesis 'A Study of Advanced
Array Processor Architectures and Their Optimisation for
Image Processing Tasks', Terry Fountain briefly examined*

some of the practical possibilities for dealing with this problem in CLIP4-like structures. The survey forms the first part of this Chapter.

The second area of development concerns the mode of control of an assembly of many processors. This area includes some of the most interesting issues in the design of large computing engines today. The system used in CLIP4 and its cousins, whereby all processors in an assembly execute the same operation at each step of the program, is well understood and easily controlled by normal sequential methods.

However, it is apparent that an alternative strategy exists, whereby each processor in an assembly can operate to some degree independently of the others. Most of the systems using this method, however, have structures with little or no natural relationship to image data. It is evident that much of the success of CLIP4-like array machines occurs because the network structure mirrors very closely that of the data involved. It is also apparent that, where such a relationship is missing, two problems exist. First, satisfactory methods for task partitioning only exist for low-level operations and, second, communication between processors quickly becomes a bottleneck for more than a few elements.

The CLIP7 project, described here by Terry Fountain, attempts to investigate these problems by means of progressive advances from the traditional array architecture, using a new custom LSI chip.

1. SCANNING IN ARRAY PROCESSORS

An array processor has, conceptually at least, the same number of processors as there are pixels in the target data field. In the case of our own target specification this would imply 262,144 processors. In order to avoid such currently impractical numbers, array processors adopt the strategy of scanning fewer processors over the data field, an idea shown conceptually in Figure 9.1. The block of processors moves to each sub-area of the data array in turn and performs the same processing on each.

There are two crucial facets to a system of this sort. Firstly, since the connectivity between processors is a vital part of an array processor, some means must be devised to ensure that the appropriate exchange of signals between sub-areas is catered for. Secondly, each processor must deal with data from some, possibly large, number of pixels, so the amount of local data storage per processor must be suitably enlarged. Given these requirements, there are at least five strategies which can be adopted.

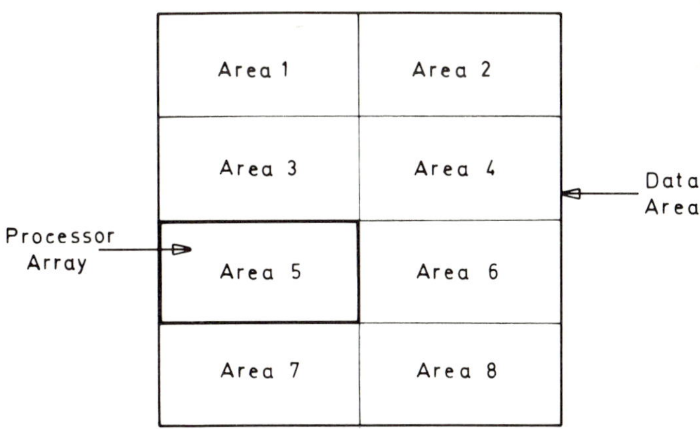

Figure 9.1   The concept of scanning.

1.1 Software Scanning

In this system, used in the Massively Parallel Processor [62], propagation signals which are to be passed between neighbouring sub-areas are stored at the edge locations of planes of local storage and are shifted across the array as required. The scheme has the dual disadvantages of consuming both time and local data storage rather excessively. The major advantages are that no extra hardware is required and the size of the target data set is limited only by the amount of local data storage available.

1.2 Overlapping Scans

Shown in Figure 9.2, this idea makes the tacit assumption that objects to be processed are no larger than some defined fraction (often half) of the array size and short-circuits the interconnection requirement by guaranteeing coverage of such objects by a coherent block of array.

The shortcomings are that the maximum size of objects is fixed and the scanning time is longer than would otherwise be necessary.

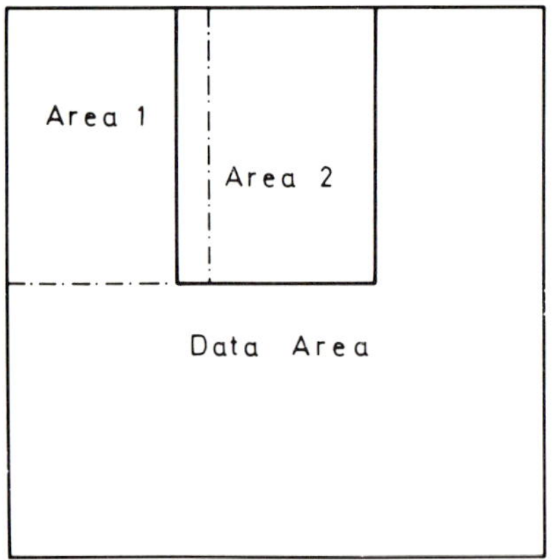

Figure 9.2   Overlapping scans.

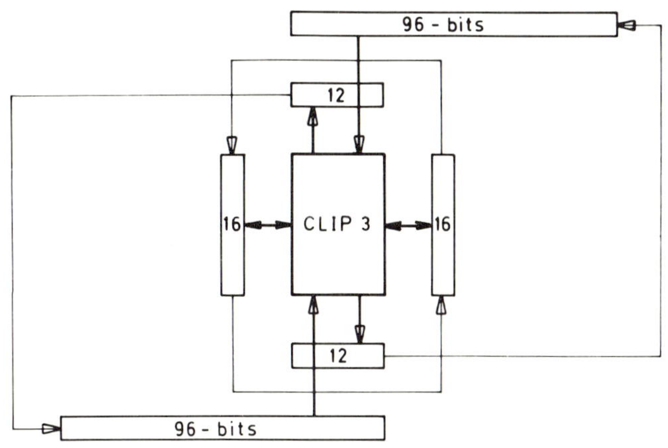

Figure 9.3   Edge store arrangements used in CLIP3/CIMU.

1.3 Two-dimensional Scanning with Edge Stores

This method, shown in Figure 9.3, was used in the CLIP3/CIMU system [63]. The special-purpose edge stores which are required along each edge of the array demand additional hardware and, furthermore, a problem exists at corner-points of the array, concerning the two possible sources for the propagation input at each corner of the array. This difficulty can be resolved by replacing the edge stores with a guard ring of dedicated processing units, but this involves further hardware complexity.

1.4 One-dimensional Scanning with Edge Stores

Figure 9.4 shows this method to have two advantages over that described above. Firstly the system implementation is simplified, since propagation signals are all of the same type and are dealt with in the same manner. Secondly the necessity for propagating across the corners of sub-arrays is removed, so there is no need even to consider the corner-point problem.

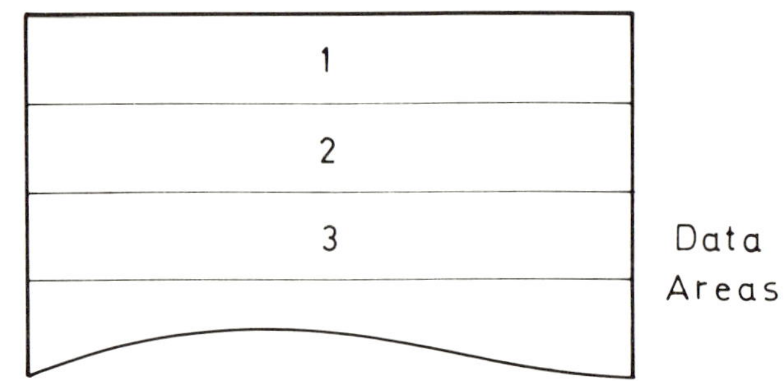

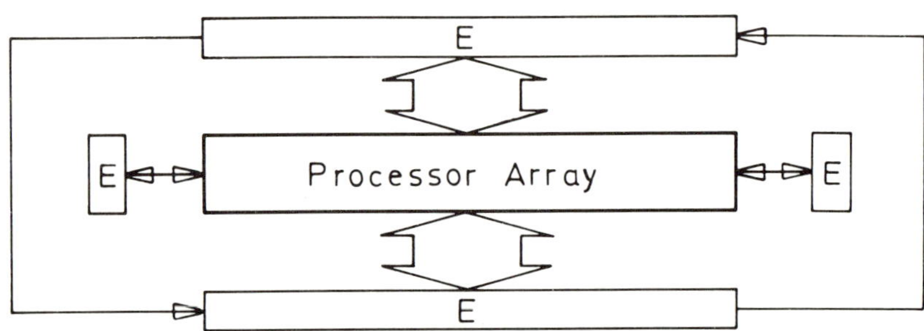

Figure 9.4   One-dimensional scanning with edge stores.

The disadvantage of this scheme is that the flexibility of choosing the area over which scanning can usefully take place is substantially reduced.   However, this can have the advantage of encouraging a user to consider the machine as a full array processor and use it accordingly.

1.5 Non-contiguous Segmentation

A scheme suggested by Danielsson and Ericsson [64] is shown in Figure 9.5. It is in a rather different class from the other methods described, in that the array of processors does not deal with a contiguous block of pixels. The concept is further complicated by the idea of offset processor/memory architecture, introduced in the same publication.   The scheme involves additional complexity of hardware and, although it offers increased efficiency, introduces practical problems of considerable severity.

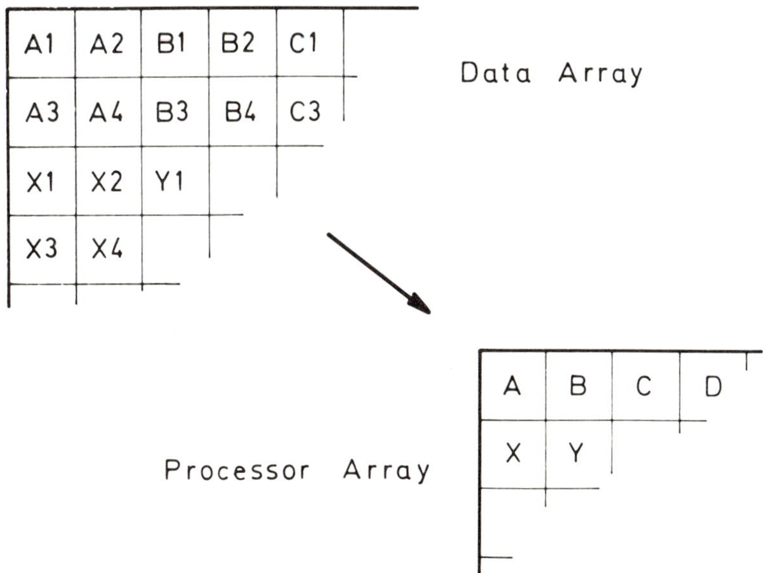

*Figure 9.5   Non-contiguous segmentation.   Each proces-
sor deals with data from similarly lettered pixels.*

2. THE CLIP7 PROJECT SPECIFICATION

The principal aims of the CLIP7 program are threefold:

1. To enable a machine of sufficient power to be
    constructed so that high-resolution (512 x 512
    pixel) images can be processed at rates compar-
    able with those at which the original CLIP4
    processes 96 x 96 pixel images.

2. To permit control strategies other than those
    whereby all processors execute the same opera-
    tion simultaneously and, in particular, to study
    the categories of problem where more individual
    processor autonomy might be useful.

3. In conjunction with the second aim, to investi-
    gate alternative system architectures, particu-
    larly three-dimensional structures having novel
    connectivity nets.

In pursuing these aims, three guiding principles have been used:

1. To provide upward compatibility of software, starting from the CLIP4 base.

2. To attempt to produce well-integrated systems, in both hardware and software, the systems not to be constrained by arbitrary categorisations.

3. To eschew advances in technology in favour of those of concept. In particular this applies to the use of leading-edge VLSI custom circuit techniques for the processors.

In order to implement the program, the following methods were adopted:

1. To devise a new LSI custom chip which can be used at all stages of the project.

2. To devise, as a first step, a system architecture which will principally fulfil the first and second project aims.

3. To construct small-scale pilot systems for any subsequent machines. This technique proved its worth in the CLIP4 program when CLIPs 1-3 provided fresh insights as a result of hands-on experience.

Accordingly, the specification shown in Table 9.1 was assembled and used to guide the design of both system and processor chip.

Table 9.1   CLIP7 system specification

Data area	512 x 512 pixels
Data source	625-line TV camera
Operating speed	As CLIP4
Processor array	512 x 4 processors
Processor	Custom LSI
Local data store	16-256 bytes/pixel 400 ns access-time RAM
Backup data store	1000 8-bit images 65 ms access-time/image

3. THE CLIP7 CHIP

The means by which this specification was developed are detailed in [65] but the purpose of this section is to present that specification, summarised in Table 9.2, as an accomplished fact. The chip data paths are shown in Figure 9.6. A general description of the manner in which the chip functions is followed by examples of the classes of operation which can be carried out by the device.

Table 9.2   CLIP7 chip specification

Technology	5 μm CMOS
Active devices/chip	5000
Die size	4.5 x 5.0 mm
Package	64-pin DIL or flatpack
Clock frequency	5 MHz
Power consumption	0.75 W
Processors/chip	1
Architecture	16-bit ALU Multi-mode shifter 8-bit parallel data accesses Serial connections to 8 neighbours 16-bit local condition register Separate local data store

3.1 Data Paths

The CLIP7 chip data paths consist essentially of two sections. The first comprises a 16-bit ALU having eighteen functions (16 Boolean, add and subtract) operating on two inputs. The first input is always derived from a 16-bit shift register (S). This register has left or right direction of shift, 16-bit arithmetic shift, 16-bit logical shift and 8-bit logical rotate modes. It also has an external data input which may be used in both left and right modes. The second ALU input can be derived from any one of five 16-bit registers (C and B0-B3). The ALU data output may be

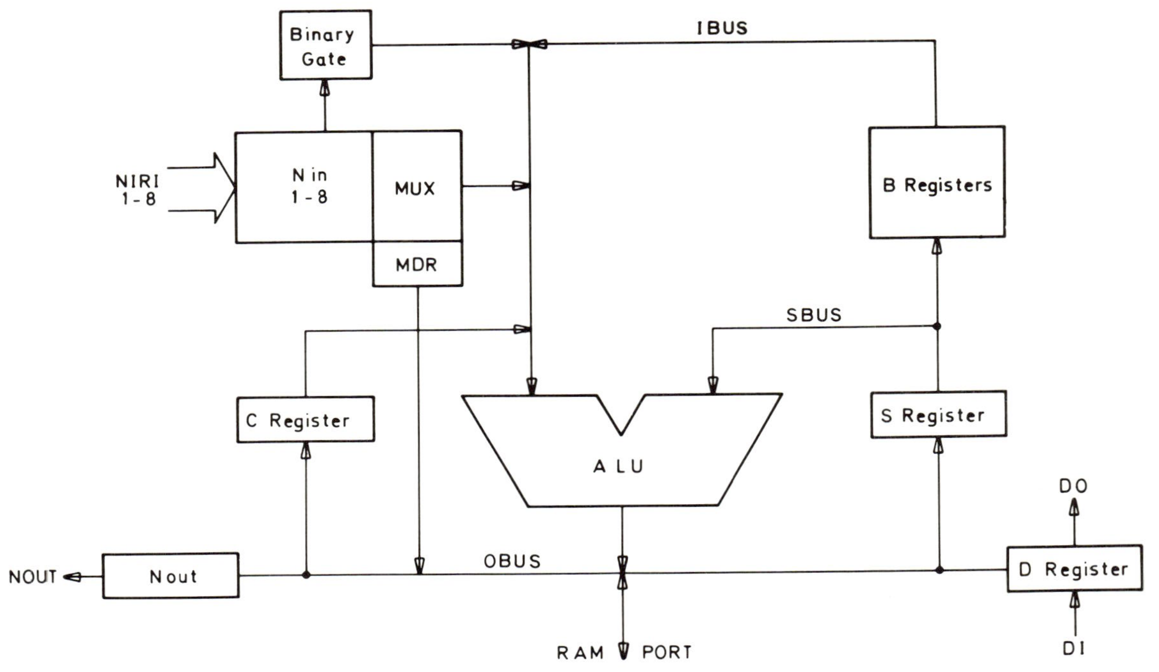

Figure 9.6   Data paths in the CLIP7 chip.

loaded to S, C, or D registers or to the bi-directional RAM access port.   Data may be input to the system via the RAM port or the D register and may be loaded to either S or C.

The second section of the data path concerns connections to neighbouring processors.   Data from the ALU outputs, D or the RAM port may be loaded to the 8-bit Nout register. In a planar array of processors, connected as shown in Figure 9.7, data is passed bit-serially to the neighbours' Nin registers.   On each chip these registers, a 64- to 8-line multiplexer and a binary gate form an input matrix. The 8-bit output of the multiplexer and the 1-bit output of the binary gate form two further alternative inputs to the ALU.

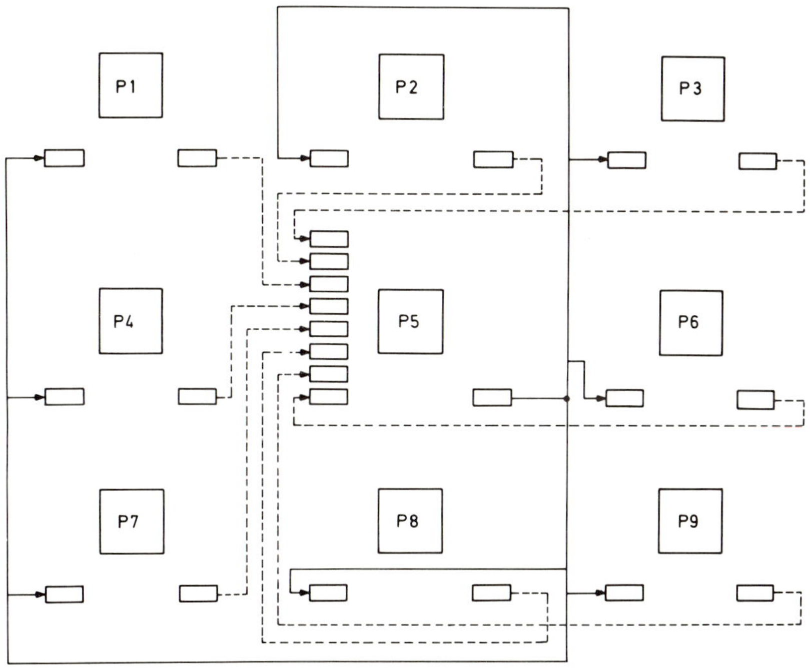

Figure 9.7   Neighbourhood connections in an array of CLIP7 chips.

The position of the MDR register within the chip structure is not easily categorised. It stores the direction of the multiplexer control at the time when the S register is loaded. The 3-bit output of the register can provide an alternative source for the loading of S, C, D and the RAM port.

3.2 External Control

The CLIP7 chip may either be driven completely by externally imposed control, or by a combination of external and internal functions. The choice is determined by the state of the 'USE CC' control pin. All operations of the circuit are fully synchronous with the 5 MHz clock.

3.3 Internal Control

Internal control is achieved by means of the C and MDR registers. The contents of these registers can affect a number of processor operations.

The MDR register records the direction from which a specific input was derived and can select a given Nin register for clearing. The register outputs can be made available to the OBUS and can thereby be used, via the C register, to control subsequent circuit operations.

The C register is concerned with four areas of processor operation: neighbourhood directionality; ALU operands; ALU operations; and register loading. It achieves control over these areas as follows:

1. The set of enabled directions for the binary gate is always controlled by the C register bits 0-7.

2. The multiplexer direction may be controlled by bits 0-2.

3. The B register address may be selected by bits 8 and 9.

4. Bit 10 may determine whether the ALU performs a PASS or an ARITHMETIC operation.

5. Bits 11 and 12 can be chosen as alternative carry inputs to the ALU.

6. Bit 15 can determine whether S, C and Nout registers are loaded.

The contents of the C register may be generated in a number of ways. They may be applied externally via the S register serial input, the RAM port or the D input. They may be generated on-chip as the result of an ALU operation. Bits 0-2 can be generated by means of an MDR output operation (other bits are cleared). Finally, bits 10, 11, 12 and 15 may be loaded during an ALU status operation (other bits are retained). In this case, the individual sources for the bits are:

Bit 10   S register bit 0 (intended for use in multiply and divide).

Bit 11   ALU bit 2 (intended for use in rounding of results).

Bit 12   ALU carry output (intended for use in multi-word operations).

Bit 15   Selected by ALU opcode from amongst ALU carry, overflow, sign and zero status bits.

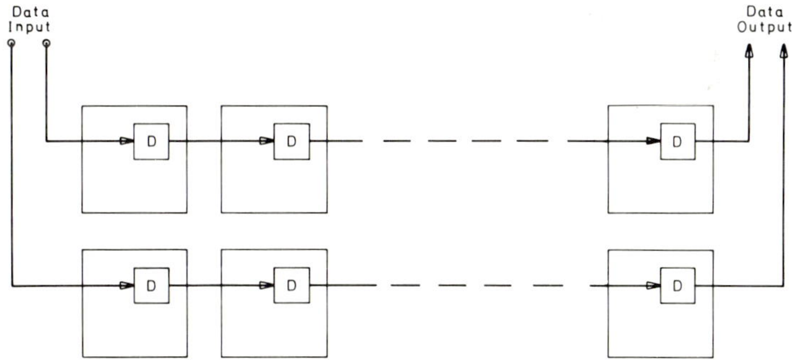

Figure 9.8   Serial data connections are made between the D registers along each row of an array of processors.

3.4 Operation and Performance

The basic modes of operation of the circuit comprise the following:

Serial data transfers   These can be either input or output operations and occur via the D registers, connected as shown in Figure 9.8, and interfaced to suitable framestores.

Local data storage and retrieval   These paired operations involve transfers between the OBUS and external storage via the RAM port.

In general, one or both of these types of operation will precede one of the following categories:

Pointwise operations   Here the intention is to produce a result which is a function of two or more inputs already resident in the local RAM.   A simple example is the addition of two numbers.   The first operand moves from RAM to S register to BO.   The second operand then moves from RAM to S register. The addition is performed in the ALU and the low-order byte of the result is loaded directly to RAM. If necessary the result is also loaded to the S register and shifted so as to allow the high-order byte to be transferred via the ALU to the OBUS and thence to RAM.

Neighbourhood operations   In operations of this type the processor derives a result which is a function of its own value and those of its neighbours. The first step is usually to transfer a byte of data from RAM to Nout and, often simultaneously, to the S register.   Data is then transferred from each cell to its neighbours, each Nout to a set of Nin registers as shown in Figure 9.7.   In general, the input multiplexer then sequences round the inputs and, in each direction, the ALU computes a new result which is usually stored in the S register.   Finally, the result is transferred byte by byte to RAM via the ALU.

In order to clarify the operation of the circuit, some specific examples of various types of function will now be given, segmented according to their radius of action in an array.   It should be emphasised at this point that the times calculated are those for a processor operating under 'one pixel per processor' conditions.

Pointwise Operations

Single operand   Examples of this type include inversion, decrement and increment.   The sequence of steps requires 4 μcycles for execution, i.e. 0.8 μs.

Double operand   Examples are addition, subtraction, multiplication, division and a variety of Boolean functions.   Operations in this category can vary widely.   A typical short sequence is addition (8 μcycles), whereas a typical long operation, floating point multiplication, uses some of the conditional facilities of the chip to improve performance and requires 108 μcycles.

Multiple operand   This type of operation often occurs as a final stage of some more complex neighbourhood function, whereby a set of relevant data has first been assembled in each processor's RAM prior to a function such as a median filter being performed.   A more strictly pointwise operation of this sort might be the addition of a number of scenes in a sequence, prior to averaging.

Local Operations

These operations generally involve a processor having access to data from its local 3 x 3 neighbourhood of pixels. Because of packaging constraints, multi-bit numbers are transferred serially.   It is worthwhile, therefore, to examine separately the 1-bit and 8-bit cases and this also serves to emphasize the different roles of the binary gate and multiplexer sections of the circuit.

One-bit   Examples of this type of operation include the use of binary masks for thinning binary images and the detection of edges in binary images.   Such operations require 8 μcycles.

Eight-bit   Examples include MAX, AVERAGE, MEDIAN and other ranking filters.   The sequence required to implement the MAX filter takes a total of 44 μcycles.

Large Radius Neighbourhood

In systems dealing with high-resolution images, many of the ranking filter operations are required to operate over much larger areas of image than the local 3 x 3 neighbourhood. Operations of this type depend for their efficient execution as much upon good data-shifting capabilities in a system as upon efficient operation of the ranking algorithm by the processor. Matthews [66] has considered the problem of large radius median filter operations on a CLIP7 system in which a strip of 512 x 4 processors is scanned over a 512 x 512 pixel image. He concludes that the scanned operation of the system is an advantage which, combined with fast shifting via the D registers, results in a data assembly time of 15 ms for a 15 x 15 window over a 512 x 512 pixel image. This process results in each processor having to perform the median filter operation on 128 sets of accumulated data, because of the scanned nature of the system. The time taken to implement the median filter algorithm for each of the 128 operations is approximately 3 ms. The algorithm uses the conditional operations of the circuit to improve performance.

Global Operations

Global operations are of considerable significance in CLIP4 for two principal reasons. First, they provide a means by which information can be propagated across a complete array, in a data-dependent fashion, by means of a single instruction. Otto describes a large set of such operations in Chapter 2. The second reason for their importance is associated with the control structure of the CLIP4 chip, because of which a nearest neighbour operation takes about 10 μs to transmit data a distance of one pixel, whereas global propagation occurs at approximately five pixels per μs, more than an order of magnitude improvement in performance (these figures refer to the CLIP4D chips).

In the CLIP7 chip this disparity is no longer so marked, partly because global propagation has become less efficient, and partly because the parallel control structure of the chip makes neighbourhood operations more efficient. The global equivalent of the local MAX filter operation (a rather inefficient and peculiar way of converting a complete image to its maximum grey level) requires 44 μcycles per pixel radius, exactly the same as that for the local MAX. Each represents approximately one bit per pixel per μs. There is no longer an exact equivalent of the CLIP4 global propagation operation, whereby the application of a single control word can result in data propagating across

an indeterminate number of cells. This reflects the
microcoded nature of the chip as well as the lack of a propagating type of function.

4. THE CLIP7 SYSTEM

Two of the principal aims of the CLIP7 system, set out
in section 2, are to provide a fast, high-resolution image
processing system and to allow investigation of the concept
of local autonomy within an array of processors. In order
to accommodate these two aims it was felt necessary to provide the system with a microcode-based control structure.

Given this decision, and the desire to produce a well-
integrated system wherein the array processor did not so
prefigure the structure as to prevent the eventual incor-
poration of other types of modules, a system based on two
buses was developed and is shown in Figure 9.9. Major components of the system, which is principally arranged to deal
with a 512 x 512 x 8-bit image format, are as follows:

Host A high-powered general-purpose computer running, under the UNIX system, all necessary user
software including editors, assemblers, compilers,
etc.

Controller An interface between host processor and
Microbus and Databus, allowing microcoded control of
all modules interfaced to these buses.

Databus A 32-bit tristate bus running at 5 MHz.

Microbus An expandable micro-control bus having
initially 128 lines.

Data I/O A structure including framestores, data
counters, host/array data interface and image capture and display facilities.

Processor array An array of 512 x 4 CLIP7
processors, together with appropriate edge stores
permitting scanning of the array in both vertical
directions.

Local storage array An array of high-density
static RAMs configured as 512 x 512 x n bytes, where
n lies between 64 and 1024 depending on the stage of
system development.

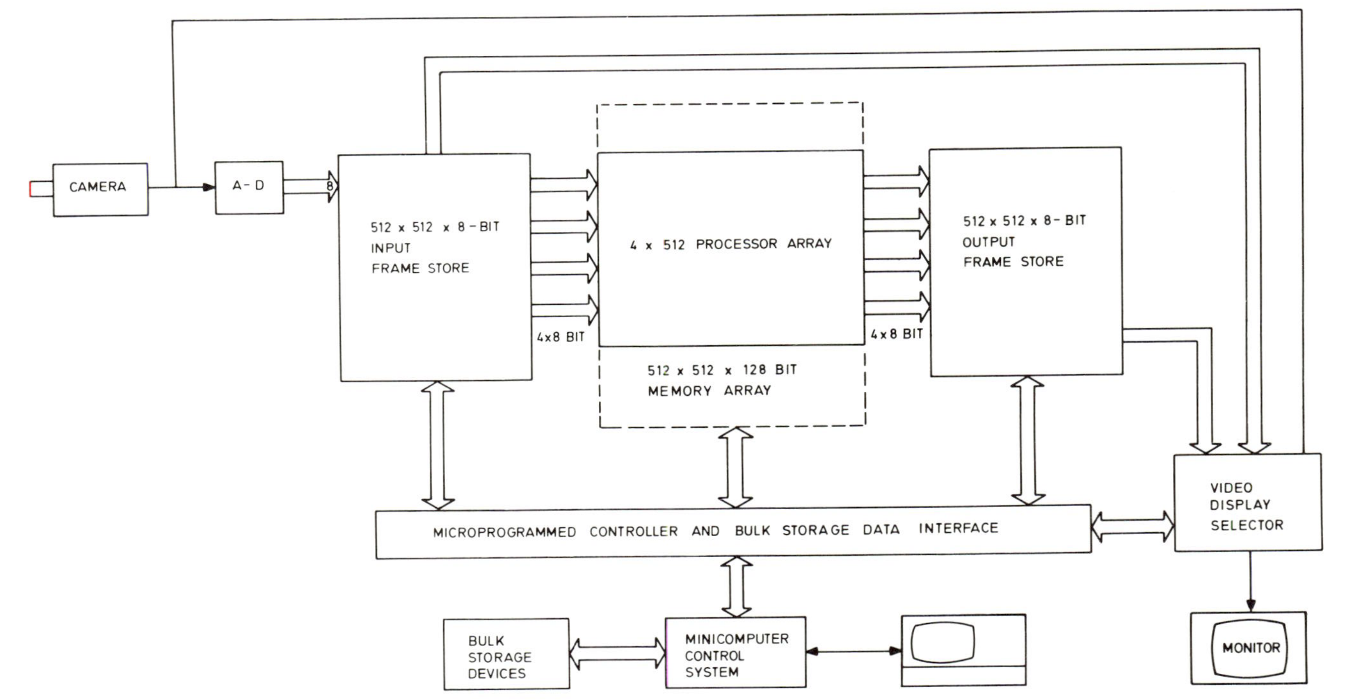

Figure 9.9   The CLIP7 system.

Array gate   An overall gate structure having a variety of functions including overall OR, position of source, send to specific destination and overall broadcast.

From the user's point of view the system may be operated in two quite distinct ways.   At one level, all the structural subtleties may be ignored and the system may be used as though it were simply a full-coverage SIMD array processor, driven solely from the high-level language IPC (see Chapter 3).   For this type of use, a library of operations, shown in Table 9.3, is provided, which parallels that currently provided for CLIP4 in the IPC subroutine library. Under these circumstances the only visible differences from CLIP4 will be the increased resolution and improved performance associated with the larger amount of local storage available.

However, software tools will also be provided to enable the system features to be utilised much more fully and, for this purpose, a more thorough understanding of the system architecture is required.   The next sections provide the necessary detail.

4.1 The Array

Under this heading are included three aspects of the system: processor array, local storage array and array gate.

The array consists of a strip of 512 x 4 processors which may be scanned over the 512 x 512 pixel data area. The size of this area may be modified by the user as described below.   This is achieved, as shown in Figure 9.10, by storing data from every fourth pixel (vertically) in the picture in the block of RAM to which each processor has access.   Thus, in order to perform a given operation over a 512 x 512 pixel image, each processor must function 128 times.   In CLIP7 each pixel of data will be provided with 64-1024 bytes of storage (dependent on the state of system development) so that each processor has access to 8-128 Kbyte of storage.

Three important points follow from this arrangement. First, it is possible to choose to scan over a smaller (or larger) vertical dimension than the usual 512 pixels and, in so doing, to vary the amount of local storage available per pixel.   Thus, if the chosen scan area is 512 x 128 pixels,

Table 9.3   Standard CLIP7 routines

Category	Operation	Category	Operation
I/O	Camera to input memory	Filter	Local median
	Camera to input memory (loop)		Large radius median
	Disc to input memory		Average
	Input memory to array		Max
	Disc to array		Min
	Count		Extremum
	Array to output memory		
	Array to disc	Spatial	Shift
	Disc to output memory		Rotate
	Output memory to disc		
	Input memory to disc	Edgeset	Constant
			Max
Boolean	16 possible		Min
			Average
Arithmetic	Add		Median
	Subtract		
	Multiply	Constant	Distribute
	Divide		
	Scale	Mask	Thicken
	Compare		Thin
	Max		Mask
	Min		Skeleton
	Absval		Rotate mask
	Square root		
	Histogram		

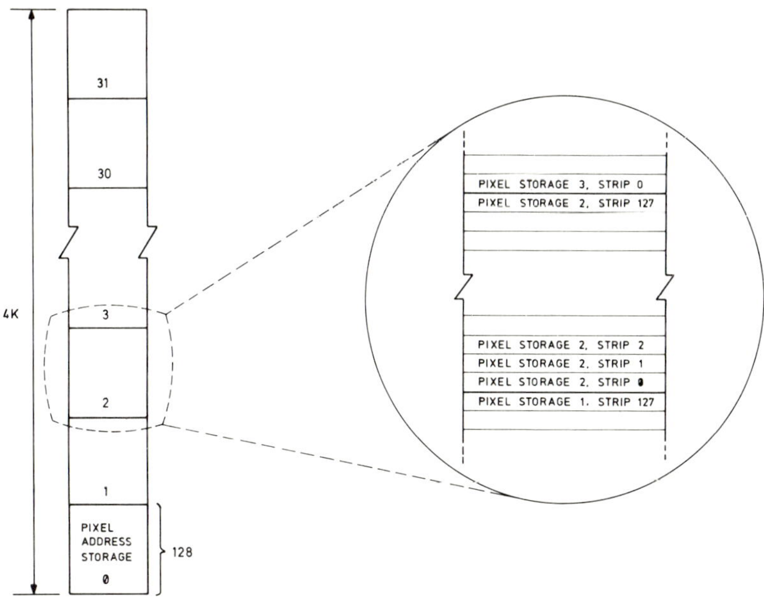

Figure 9.10   Partitioning of the local data memory.

the amount of local storage available per pixel is
quadrupled.   The second point concerns the possibility,
permitted by the control system, of assigning a block of the
RAM available to each processor as an overwritable or re-
accessible workspace.   As an example of this, if each pixel
normally has available 64 bytes of RAM (8 Kbyte per processor) then a workspace of 128 bytes can be assigned to the
processor, whilst the amount of storage available per pixel
is reduced to 63 bytes.   The third point concerns data
shifting.   By manipulation of the RAM addresses both
apparent and actual shifting of data in the vertical direction can be achieved at high speed.

The array gate forms an important addition to the array.
In the CLIP4 system an array gate was provided which formed
an overall OR function of processor outputs.   This was used
principally to detect the 'array clear' condition, and the
CLIP7 array gate still offers this facility.   However, its
operation has been enhanced so that the following additional
facilities are available [67].

First, it allows the position of the first (top leftmost) non-zero array point to be read directly. Second, the value of data at that point can be retrieved directly. Third, a value can be passed from the controller to any specified processor in the array. Finally, a constant can be broadcast to all processors in the array simultaneously. In these two latter operations, data is fed to the serial inputs of the S registers of the processors concerned.

During a scan of an image, the array clear/not clear output can be used to pinpoint (via control registers) those points in the scan where a relevant image property began and ended. Used in conjunction with the properties of the gate structure described above, these control registers can yield positional information about the array for use in subsequent operations or by the host processor.

4.2 Data I/O

This section of the system includes the Databus, framestores, counters, host/array data interface and image capture and display facilities. The principal design aim of this part of the system was to allow for flexible expansion of facilities. It was decided to achieve this by specifying a common system databus to which each section of the system dealing with data is interfaced. The Databus is a 32-bit wide, 5 MHz tristate-driven bus to which one driver and any number of receivers may be connected at any one time. This structure allows direct transfers, for example, from output framestore to input framestore, or from array output to array input. (This second example offers an interesting extra possibility for data shifting in the array.) The Databus itself involves no handshaking control; all transfers are directly governed by the Microbus (see below). The 32-bit configuration is obviously derived from the processor array's four 8-bit serial data connections, and other components of the system are designed to reflect this format.

The data capture system comprises a normal monochrome CCTV running in interlaced mode (i.e. 40 ms per full-frame capture time), high-speed 8-bit A/D converter and input framestore. All data transfer operations in the I/O system are defined to be of 8-bit data, so no bit selection or thresholding facilities on the framestore outputs are offered. The output framestore and input framestore are similarly configured, and false colour as well as monochrome display facilities are provided.   All strictly display facilities are directly under host control (via a register interface).

In CLIP4 a counter was provided which produced a sum of all bits set in the array.   For CLIP7 this is generalised to act on 8-bit data from each processor.   However, the CLIP7 array itself can be utilised to perform a part of the counting operation.   Thus each processor can be used to add directly data from the 128 pixels to which it has access, forming a word-length sum.   The hardware counter is then required to add these 2048 words, forming first the sum of the low-order bytes then adding this to the sum of the high-order bytes, in the time taken to serially output two strips of array data.   The maximum sum derived from this operation is 26 bits long, and therefore the result of a count operation is a two-word sum.   Even the maximum potential sum of a 1-bit 512 x 512 image requires a double word.

Data transfers between array and host are handled by an interface which is essentially another framestore.   This means that the image size is predefined at 512 x 512 bytes, and also that transfer operations require time for both loading and unloading the framestore.

All the facilities initially provided by the I/O assume the 512 x 512 pixel format but, just as the array can be made to scan over a variety of different areas, so could the Databus and control system handle a variety of data formats if suitable modules were to be configured.

4.3 Control

The system controller comprises the host processor, the CLIP7 control unit and the Microbus.

The host processor provides the environment for all system software (editors, compilers, subroutine libraries, etc.) with the preferred operating system being UNIX.   Programs, using standard subroutines, are written in IPC and the available subroutine library is such that the great majority of CLIP4 programs will be upward compatible.   Support software will be provided for the construction of new microcode routines at two levels.   At the lower level, complete lines of microcode can be constructed and assembled into routines.   This has the potential for maximum efficiency.   At the intermediate level, preconfigured lines or short segments of microcode which correspond to assembly-level instructions can be assembled with relative ease into routines of moderate efficiency.

The host processor also performs the second function of driving the CLIP7 system by passing appropriate instructions to the CLIP7 control unit and receiving from it status information, data and interrupts.

The CLIP7 control unit, shown in Figure 9.11, consists of two functional parts, the microprogram sequencer and the microprogrammed controller. The sequencer generates the control signals for nearly all of the CLIP7 system by sequential access to a large (128 bits by up to 64 Kword) writeable microcode instruction store. The generation of the addresses supplied to this store is performed by the micro-code sequencer controller (Am 29112) which is capable of sequential addresses, as well as direct, relative and sub-routine jumps dependent on the status of the system. The sequencer controller includes a 32-word stack and two 8-bit counters which are used to control microcode loops and may be reconfigured as a single 16-bit counter. The sequencer is also interruptible in both maskable and unmaskable modes, the latter allowing error recovery on stack overflow or underflow.

The system controller is responsible for host computer interface as well as the generation of the instruction operand dependent control signals, such as RAM addresses and direction selections. It consists of a block of memory of up to 64 Kword, sets of array control registers, Databus control registers, host interface registers and a microprogrammable processor (Am 29116) which includes 32 registers, an accumulator, a status register and an ALU.

The most important of the array control registers is the array local storage address register, which defines the address of the data held in local storage which is accessed by the CLIP7 processors. In order to clarify the use of this register, consider the operation of adding two 8-bit, 512 x 512 images together. The instruction, at its simplest, will consist of the addition opcode and two image addresses, defining the images to be added, and a third des-tination address. The arrangement of the images in the local storage of the array is such that images form a 128-byte block, the base address of which is given by the instruction image address field. The routine must add the position of the processor array across the image as it scans

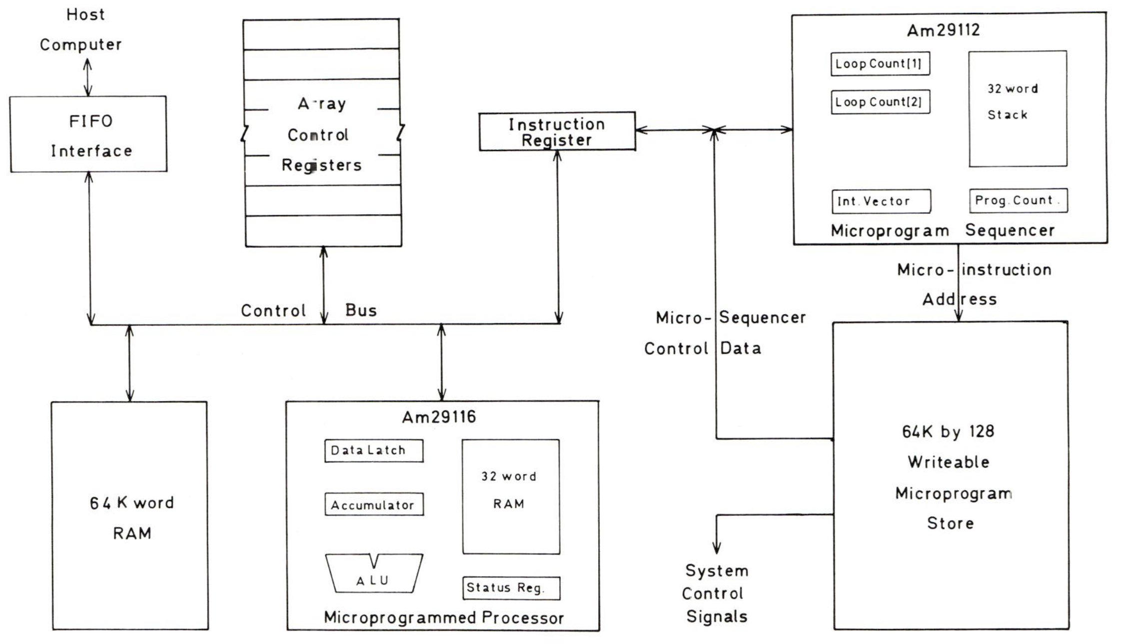

Figure 9.11   The CLIP7 control unit.

to the base address to give the required local storage
address. The microprogrammed controller is designed in such
a way as to allow independent operations to occur simultaneously.

The remaining array control registers are the serial I/O
register, the write position register, the read position
register and the array MUX control register.

The CLIP7 system allows two modes of writing data to the
processor array, either singly to a processor whose position
is specified by the address in the write position register,
or globally to the entire array. The data to be written is
loaded into the serial I/O register and is passed serially
to the S registers in the processor array.

Various status parameters of the array are accessible to
the controller. Firstly, there is the normal overall
gating which forms the logical OR of the most significant
bit of the RAM port (bit 7) of each of the processors.
This overall gating is available as the control status
register bit 0, as well as being available as an external
control signal to the microcode sequencer controller, allowing such operations as 'jump if array overflow set' or
'return from subroutine if array not zero'. Finally, the
overall gating is used as the serial input to the array
serial I/O shift register which allows the data from a single processor in the array to be read in a straightforward
way.

Associated with the overall gating is the array read
position register which indicates the position of the first
processor within the array whose status output signal (RAM
port bit 7) is not set. First processor means the top,
leftmost processor in the array, i.e. priority is given from
left to right over the priority from top to bottom.

Finally, the array MUX register selects the Nin register
to be used in the micro-operation and allows instructions
such as image shifting to include a direction list and hence
reduce the quantity of microcode to be written. The register may be cleared and incremented allowing simple and rapid
scanning of all eight nearest neighbour directions.

Because of the scanned nature of CLIP7, the average
instruction execution time is relatively long, which allows
preprocessing of the instruction queue before execution to
be performed without undue overhead. This preprocessing
comes in the form of instruction format verification, i.e.
checking the number and type of operands supplied with the
instruction, and memory management of the operands if

required. This memory management allows virtual addressing
of up to 1 gigasegment, where a segment may be of any size
from 1 byte up to 32 Kbyte (maximum local storage size).

5. SYSTEM PERFORMANCE

System performance as perceived by a user is compounded
of a number of facets, some easily quantified, others less
so.   The first of these concerns the available software.
The CLIP7 system provides essentially three levels of
software, of which the highest is IPC, incorporating a set
of standard operations found to be useful on CLIP4.   At the
next level, users can write routines in the CLIP7 assembly
language.   This level requires an understanding of the
operations performed by both system and chip, but not of the
means by which these actions are achieved.   At the lowest
level, routines can be written in true microcode, which
naturally requires the deepest understanding.

The second aspect of system performance is connected
with the efficiency of coupling between processor and host.
The dual bus structure of the CLIP7 system, in conjunction
with the two-channel instruction pipeline, means that true
overlapped fetch and execution of instructions can be
achieved.   Since, in CLIP7, the execution is usually the
longer operation, the processor is usually run at maximum
efficiency.   This situation may be different, however, if
very short, non-scanned segments of microcode are being
used, and this factor should be taken into account by potential users of the system.

The third aspect of performance, the raw operating power
of the processor, is easiest to quantify and is summarised
in Table 9.4.   However, it should be noted that these times
refer to relatively low-level, fully scanned standard operations, and it is expected that the unusual facets of both
chip and system will yield better performance for complete
programs once they have been investigated and exploited.
Aspects of the machine which may be of significance are the
local autonomy of the chip, the ability to easily rotate
local connectivity masks, the flexibility of the addressing
modes of the local memory, and the ability of the system to
concentrate attention on less than the full image.

Table 9.4   Processor operating times

Operation	Time in ms
Binary pointwise logical	0.2
8-bit pointwise add	0.2
8-bit pointwise multiply	1.9
Binary neighbour mask	0.2
8-bit neighbour maximum	1.2
8-bit neighbour median	5.1

A fourth important aspect of system performance concerns the manipulation (input, storing, fetching and display) of images. A single 512 x 512 x 8-bit image comprises 256 Kbyte of information. To transmit one such image at the normal bus transfer rate of (say) 1 Mbyte/s requires about 250 ms. Normal video data transfer rate is about 13 Mbyte/s and the capture or display of one full image requires 40 ms. The CLIP7 system provides local storage for up to 1024 images with an access time of 50 µs per image (incidentally requiring 256 Mbyte of solid-state memory to do so).

One final constraint on system performance which needs to be considered is the possible use of CLIP7 in a timeshared environment. The difficulties here principally concern the amount of image store available, although any specialised user microcode would be a subsidiary problem. The detrimental effects on performance are hard to quantify and an attempt has been made to avoid them by providing in each case the maximum practicable amount of storage.

CONCLUSION

Cellular Logic Image Processing began to appear as a new concept in the literature a little more than a quarter of a century ago. At the start, the ideas tentatively being put forward seemed strange and impracticable, without a firm theoretical foundation and lacking the powerful technological support needed to translate them into computing machines. Indeed, even the phrase 'parallel processing' was largely unused within the computing community. Today, however, we see a very different picture in which many special-purpose computer architectures have been implemented in software, although rather less in hardware, a respectable number being built into complete systems commissioned and applied in image processing laboratories.

The CLIP programme was only one of several which emerged at a time when interest in vision began to stimulate research into automatic image analysis. It could be argued that the degree of success attained by the processor array approach to image computing has been largely due to the similarity, tenuous though it may be, between processor arrays and the retina and visual cortex. Whether or not this is true, similar arguments might also be used to explain why much of the success has been concentrated on low-level processing. Just as high-level vision is known not to be located within the parts of the brain logically near to the retina, so it would seem that automatic high-level vision processes do not map naturally on to CLIP4 and similar architectures. In the CLIP4 system, the host serial computer works in close collaboration with the array and can often take on the higher-level processing for which it is still competitively efficient, leaving the lower-level processing to the array. Occasionally, a careful rethinking of the high-level tasks will suggest algorithms which run efficiently in the CLIP4 environment but the amount of ingenuity demanded seems suspiciously large: processor arrays are probably not the optimum structures for high-level image processing. This does not mean that the higher-level processes must necessarily be taken out of the array structure and, indeed, it may well be that the process of switching the computation to the serial host might in itself be more time-consuming than the inadequacies of the array.

With many of the low-level processes now being handled competently in systems such as CLIP4, it is pertinent to ask what new structures will be devised to handle the higher-level processes with comparable efficiency. Furthermore, having produced these new structures, the problem of linking the high-level and low-level machines effectively will have to be addressed.

The Image Processing Group at University College London will continue to explore the potential of cellular logic image processing and the CLIP7 project provides a focus for the research that will be carried out. Nevertheless, it will be surprising if the next quarter of a century does not produce radically different computer architectures aimed at high-level vision; it will be gratifying if the study of CLIP7 helps these to emerge.

REFERENCES

[1] M J B Duff (1962) The Digiscat — a digitally controlled device for the measurement of multiple scattering in nuclear emulsions. Nucl. Instr. and Meth. **15**, 87–94.

[2] S H Unger (1958) A computer oriented towards spatial problems. Proc. IRE **46**, 1744–1750.

[3] S H Unger (1959) Pattern detection and recognition. Proc. IRE **47**, 1737–1752.

[4] J Y Lettvin, H R Maturana, W S McCulloch and W H Pitts (1959) What the frog's eye tells the frog's brain. Proc. IRE **47**, 1940–1951.

[5] D H Hubel and T N Wiesel (1962) Receptive fields, binocular interaction and functional architecture in the cat's visual cortex. J. Physiol. **160**, 106–154.

[6] M B Herscher and T P Kelley (1963) Functional electronic model of the frog retina. IEEE Trans. **MIL–7**, 98–103.

[7] D L Slotnick, W C Borck and R C McReynolds (1962) The Solomon computer. Proc. AFIPS Fall Computer Conference, pp. 87–107.

[8] B H McCormick (1963) The Illinois pattern recognition computer — ILLIAC III. IEEE Trans. **EC–12**, 791–813.

[9] M J B Duff, B M Jones and L J Townsend (1967) Parallel processing pattern recognition system UCPR1. Nucl. Instr. and Meth. **52**, 284–288.

[10] S Levialdi (1968) CLOPAN: a closed-pattern analyser. Proc. IEE **115**, 879–880.

[11] M J B Duff and D M Watson (1971) Automatic design of pattern recognition networks. Proc. Electro-Optics '71 Int. Conf., Brighton, England. Industrial and Scientific Conference Management Inc., Chicago, Ill., pp. 369–377.

[12] D M Watson (1974) The Application of Cellular Logic to Image Processing. PhD Thesis, University of London.

[13] J Serra (1980) The Boolean model and random sets. Comp. Graph. Image Proc. **12**, 99-126.

[14] M J E Golay (1969) Hexagonal parallel pattern transformations. IEEE Trans. **C-18**, 733-740.

[15] D Lewin (1972) The arithmetic unit. Chapter 5 in **Theory and Design of Digital Computers.** Nelson, Walton-on-Thames, Surrey.

[16] A V Aho, J E Hopcroft and J D Ullman (1974) **The Design and Analysis of Computer Algorithms.** Addison-Wesley, Reading, Mass.

[17] D E Reynolds (1983) Automatic Generation of Image Segmentation Procedures for a Cellular Array. PhD Thesis, University of London.

[18] S Levialdi (1972) On shrinking binary picture patterns. Comm. ACM **15**, 7-10.

[19] C V Kameswara Rao, B Prasada and K R Sarma (1976) A parallel shrinking algorithm for binary patterns. Comp. Graph. Image Proc. **5**, 265-270.

[20] A Haas, G Matheron and J Serra (1967) Morphologie mathématique et granulométries en place. Ann. Mines **11**, 735-753.

[21] A Haas, G Matheron and J Serra (1967) Morphologie mathématique et granulométries en place (2e partie). Ann. Mines **12**, 767-782.

[22] J Serra (1982) **Image Analysis and Mathematical Morphology.** Academic Press, London.

[23] G P Otto (1984) Algorithms for Image Processing on the CLIP4 Cellular Array Processor. PhD Thesis, University of London.

[24] A M Wood (1977) CAP4 Programmer's Manual. Internal Report, Image Processing Group, Department of Physics and Astronomy, University College London.

[25] W A Perkins (1980) Area segmentation of images using edge points. IEEE Trans. **PAMI-2**, 8-15.

[26] T J Biscoe (1971) Carotid body: structure and function. Physiol. Rev. **51**, 437-481.

[27] W A Gaunt (1971)   **Microreconstruction.**   Pitman Medical, London.

[28] K A Clarke and H H-S Ip (1982)   A parallel implementation of geometric transformations.   Pattern Recogn. Lett. **1**, 51-53.

[29] A Rosenfeld and A C Kak (1976)   **Digital Picture Processing.**   Academic Press, New York.

[30] H H-S Ip (1983)   Automatic Detection and Reconstruction of Three-dimensional Objects Using a Cellular Array Processor.   PhD Thesis, University of London.

[31] D J Potter (1984)   Analysis of Images Containing Blob-like Structures Using an Array Processor.   PhD Thesis, University of London.

[32] H Fuchs, Z M Kedem and S P Uselton (1977)   Optimal surface reconstruction from planar contours.   Comm. ACM **20**, 693-702.

[33] H N Christiansen and T W Sederberg (1978)   Conversion of complex contour line definitions into polygonal element mosaics.   Comput. Graphics **12**, 187-192.

[34] D Metcalf and M A S Moore (1971)   **Haemopoietic Cells.** North-Holland, Amsterdam.

[35] P Baines, S J L Bol and M Rosendaal (1982)   Physical and kinetic properties of haemopoietic progenitor cell populations from mouse marrow detected in five different assay systems.   Leukaemia Res. **6**, 81-88.

[36] T W Ridler and S Calvard (1978)   Picture thresholding using an iterative selection method.   IEEE Trans. **SMC-8**, 630-632.

[37] J Adam, B M Wilson and M Rosendaal (1983)   The response of colony-forming cells of Balb/c mice to immunisation and infection by <u>Salmonella typhimurium</u>. Autumn Meeting of the British Society of Immunology, Communication no. 70, 11 November.

[38] E O Brigham (1974)   **The Fast Fourier Transform.** Prentice-Hall Inc, Englewood Cliffs, New Jersey.

[39] C F R Weiman (1976)   Highly parallel digitized geometric transformations without matrix multiplication.   Proc. Int. Joint Conf. on Parallel Processing, pp. 1-10.

[40] G T Herman (1975) A relaxation method for reconstructing objects from noisy X-rays. Mathemat. Prog. **8**, 1-19.

[41] G N Ramachandran and A V Lakshminarayanan (1971) Three-dimensional reconstruction from radiographs and electron micrographs: application of convolution instead of Fourier transforms. Proc. Nat. Acad. Sci. USA **68**, 2236-2240.

[42] G T Gullberg and T F Budinger (1981) The use of filtering methods to compensate for constant attenuation in single-photon emission computed tomography. IEEE Trans. **BME-28**, 142-157.

[43] G T Herman, A Lent and P H Lutz (1978) Relaxation methods for image reconstruction. Comm. ACM **21**, 152-158.

[44] L A Shepp and B F Logan (1974) The Fourier reconstruction of a head section. IEEE Trans. **NS-21**, 21-43.

[45] A M Wood (1981) The interaction between hardware, software and algorithms. In **Languages and Architectures for Image Processing**, ed. M J B Duff and S Levialdi, Academic Press, London, pp. 1-11.

[46] J K Aggarwal and R O Duda (1975) Computer analysis of moving polygonal images. IEEE Trans. **C-24**, 966-976.

[47] J W Roach and J K Aggarwal (1980) Determining the movement of objects from a sequence of images. IEEE Trans. **PAMI-2**, 554-562.

[48] C L Fennema and W B Thompson (1979) Velocity determination in scenes containing several moving objects. Comp. Graph. Image Proc. **9**, 301-315.

[49] J L Potter (1977) Scene segmentation using motion information. Comp. Graph. Image Proc. **6**, 558-581.

[50] S Ullman (1979) **The Interpretation of Visual Motion.** MIT Press, Cambridge, Mass.

[51] H-H Nagel (1978) Formation of an object concept by analysis of systematic time variations in the optically perceptible environment. Comp. Graph. Image Proc. **7**, 149-194.

[52] A M Wood (1983) Parallel Processing Techniques for Image Sequence Analysis. PhD Thesis, University of London.

[53] C Berge (1973) **Graphs and Hypergraphs.** North-Holland, Amsterdam.

[54] V Strassen (1969) Gaussian elimination is not optimal. Num. Math. **13**, 354-356.

[55] R Klette (1978) Fast matrix multiplication by Boolean RAM in linear storage. Proc. 7th Symp. on Mathematical Foundations of Computer Science, ed. J Winkowski, Springer-Verlag, Berlin, pp. 308-314.

[56] N S Prywes (1977) Automatic generation of computer programs. Advances in Computers **16**, 58-121.

[57] H Tennant (1981) **Natural Language Processing.** Petrocelli, New York.

[58] P J Hayes (1979) The naive physics manifesto. In **Expert Systems in the Micro-electronic Age,** ed. D Michie, Edinburgh University Press, Edinburgh, pp. 168-201.

[59] S D Pass (1981) Parallel Techniques for High Level Image Segmentation Using the CLIP4 Computer. PhD Thesis, University of London.

[60] D A Waterman and F Hayes-Roth (1978) **Pattern-Directed Inference Systems.** Academic Press, New York.

[61] C Lantuéjoul (1982) Geodesic segmentation. In **Multicomputers and Image Processing,** ed. K Preston, Jr and L Uhr, Academic Press, New York, pp. 111-124.

[62] J L Potter (1981) Continuous image processing on the MPP. IEEE Computer Society Workshop on Computer Architecture for Pattern Analysis and Image Database Management, Hot Springs, Va., pp. 51-56.

[63] M J B Duff (1977) Geometrical analysis of image parts. In **Digital Image Processing and Analysis,** ed. J C Simon and A Rosenfeld, Noordhoff, Leyden, pp. 101-124.

[64] P-E Danielsson and T Ericsson (1982) Suggestions for an image processor array. Internal Report LiTH-ISY-I-0507, Department of Electrical Engineering, Linköping University, Sweden.

[65] T J Fountain (1982) The CLIP7 array processor micro-circuit. Internal Report No. 82/8, Image Processing Group, Department of Physics and Astronomy, University College London.

[66] K N Matthews (1983)   Large window median filtering on CLIP7.   Pattern Recogn. Lett. **1**, 341–346.

[67] K N Matthews and T J Fountain (1984)   Apparatus and Methods for Processing an Array of Items of Data.   UK Patent Application 8429599 (23 November).